CHEMICAL MODIFICATION OF ENZYMES:
Active Site Studies

Ellis Horwood books in the
BIOLOGICAL SCIENCES
General Editor: Dr ALAN WISEMAN, University of Surrey
Series in
BIOCHEMISTRY AND BIOTECHNOLOGY
Series Editor: Dr ALAN WISEMAN, Senior Lecturer in the Division of
Biochemistry, University of Surrey

* In preparation

CHEMICAL MODIFICATION OF ENZYMES:
Active Site Studies

Editor:
JAIME EYZAGUIRRE, Ph.D.
Professor of Biochemistry
Universidad Católica de Chile, Santiago

Authors:
Sergio Bazaes, Ph.D., Emilio Cardemil, Ph.D.,
Jorge Churchich, Ph.D., Hilda Cid, Ph.D.,
Jaime Eyzaguirre, Ph.D., Robert Kemp, Ph.D.,
Thomas Nowak, Ph.D., Hans-Jochen Schäfer, Ph.D.,
Eduardo Silva, Ph.D. and John Wilson, Ph.D.

ELLIS HORWOOD LIMITED
Publishers · Chichester

Halsted Press: a division of
JOHN WILEY & SONS
New York · Chichester · Brisbane · Toronto

First published in 1987 by
ELLIS HORWOOD LIMITED
Market Cross House, Cooper Street,
Chichester, West Sussex, PO19 1EB, England
The publisher's colophon is reproduced from James Gillison's drawing of the ancient Market Cross, Chichester.

Distributors:

Australia and New Zealand:
JACARANDA WILEY LIMITED
GPO Box 859, Brisbane, Queensland 4001, Australia
Canada:
JOHN WILEY & SONS CANADA LIMITED
22 Worcester Road, Rexdale, Ontario, Canada
Europe and Africa:
JOHN WILEY & SONS LIMITED
Baffins Lane, Chichester, West Sussex, England
North and South America and the rest of the world:
Halsted Press: a division of
JOHN WILEY & SONS
605 Third Avenue, New York, NY 10158, USA

QP
601
C42
.1987

© 1987 J. Eyzaguirre/Ellis Horwood Limited

British Library Cataloguing in Publication Data
Chemical modification of enzymes: active site studies. —
(Ellis Horwood series in biochemistry and biotechnology)
1. Enzymes   2. Binding sites (Biochemistry)
I. Eyzaguirre, J.
574.19'25   QP601

Library of Congress Card No. 86–21328

ISBN 0–7458–0023–8 (Ellis Horwood Limited)
ISBN 0–470–20763–9 (Halsted Press)

Phototypeset in Times by Ellis Horwood Limited.
Printed in Great Britain by R. J. Acford, Chichester.

Table of Contents

6   **Table of contents**

Preface

This book is an outgrowth of a course on chemical modification of enzymes that was offered, under the sponsorship of UNESCO and the Alexander von Humboldt Foundation, at the Catholic University of Chile and the University of Santiago de Chile in November of 1984. Like the course, the book is mainly oriented towards graduate students and researchers interested in the study of the active site of enzymes. It is neither intended as an in-depth review of the subjects covered nor as an exhaustive revision of the literature. The main objective of this work is to present the principles behind many important approaches currently utilized in active site studies, while stressing the advantages and limitations of the different techniques. A good number of up to date references are included, however, to guide those interested in further reading.

As in the course, the main topic of the book is chemical modification of enzymes, to which the first seven chapters are devoted. In the first chapter I discuss the general principles of chemical modification and the use of group-specific reagents. This is followed by an analysis of the usefulness of kinetic techniques in chemical modification (Emilio Cardemil), and then by two chapters on the use of affinity labels. Sergio Bazaes discusses these probes, using the dialdehyde derivatives of nucleotides as examples, while Hans-Jochen Schäfer looks into photoaffinity labels and photoaffinity crosslinking. The following two chapters are devoted to specific applications of chemical modification. Eduardo Silva discusses sensitized photo-oxidation, while Jorge Churchich looks at the use of fluorescent compounds as active-site markers. Finally, Robert Kemp presents a discussion of how chemical modification can be used to learn about the structure of allosteric sites.

The rest of the book deals with physical techniques which have found wide applicability in active-site studies. Of paramount importance is X-ray diffraction, introduced by Hilda Cid. Thomas Nowak discusses how nuclear magnetic resonance can give valuable information when studying an enzyme active site. John Wilson presents an introduction to the use of monoclonal antibodies, which are being used increasingly for learning about the three-dimensional structure of proteins. Our rapidly expanding knowledge of the amino acid sequence of proteins has led to the proposal of different methods to predict secondary structures. How these predictive methods can help us learn about active-site structure is analyzed by Hilda Cid.

I wish to thank all the authors for their efforts and enthusiasm in preparing their chapters. All of us extend our appreciation to our Publisher, Ellis Horwood, particularly to Sue Horwood for her patience, help, and understanding in the different steps of the preparation of the manuscript. Finally a special word of thanks to the University of Notre Dame for the kind hospitality I received during my sabbatical year, when most of the editorial work was performed.

Jaime Eyzaguirre
Notre Dame, Indiana, August 1986

1

Chemical modification of enzymes — an overview. The use of group-specific reagents

Dr Jaime Eyzaguirre, Laboratorio de Bioquímica, Universidad Católica de Chile, Casilla 114-D, Santiago, Chile

INTRODUCTION

Enzymes represent a most remarkable set of biomolecules, owing both to their high catalytic activity and their ligand specificity. One of the important goals of biochemistry is to learn how enzymes perform their task, which is fundamental for the very existence of life.

The study of enzyme mechanisms can follow several different approaches. One of them is to analyze the kinetics of the reaction catalyzed by a certain enzyme. By this means one can learn about the order in which substrates bind and products leave the enzyme, and the velocity at which these processes take place. This analysis does not tell us, however, the detailed molecular mechanism by which these transformations occur, and how the structural elements of the enzyme contribute to them.

Enzyme and substrates come in close contact in a limited area of the enzyme surface called the 'active site', at which the catalytic process takes place. To learn about the catalytic mechanism of an enzyme it is, therefore, essential to study the structural elements of this site (amino acid side-chains and prosthetic groups) and the three-dimensional conformation of the site. The most powerful technique currently available for such a study is X-ray diffraction. If a diffraction pattern of sufficient resolution can be obtained, one can get a very accurate picture of the three-dimensional structure of the enzyme, and thus of its active site. This technique has been applied very successfully to many enzymes, some of the best known examples being lysozyme (Ford *et al.* 1974) and carboxypeptidase A (Quiocho & Lipscomb 1971). With a good knowledge of the structural elements of the active site, a reasonable mechanism for their participation in the catalytic process can be deduced.

However this... [X-ray diffraction, however, suffers from some limitations. Among them are the need of obtaining the enzyme in a crystalline form adequate for analysis. The technique is also expensive, time-consuming and requires very specialized personnel and equipment] Besides, the question still remains as to how valid are results obtained in the crystalline state when extrapolated to aqueous solution (the condition under which enzymes operate). Dr Cid discusses X-ray diffraction in greater detail in Chapter 9 of this volume.

A host of other analytical techniques, individually less powerful than X-ray diffraction, have been developed through the years to address structure questions. These techniques are important, not only as means of complementing or confirming X-ray data, but can also give information of their own. Very often these are the only ways of probing active-site structure and function. [Many of these techniques are discussed in this and subsequent chapters: chemical modification, nuclear magnetic resonance, monoclonal antibodies], etc.

CHEMICAL MODIFICATION—GENERAL ASPECTS

[Of special importance in probing active-site structure is chemical modification. By this means, a chemical reagent is placed in contact with the enzyme and a chemical reaction occurs. This reagent will bind covalently to amino acid side-chains in the enzyme and will produce changes in some measurable property (or properties) of the enzyme. Ideally, the reagent should be of sufficient selectivity to combine with only one residue, and cause minimal alterations in the conformation of the enzyme. This covalent derivatization is then correlated to the enzyme property under consideration, so that a function can be proposed for the modified residue.]

Of the 20 natural amino acids, only those possessing a polar side-chain are normally the object of chemical modification. The chemical reactivity of these groups is basically a function of their nucleophilicity. This nucleophilicity can be influenced in turn by several factors which, as will be discussed below, can greatly affect the outcome of chemical modification.

Two main types of approach to chemical modification can be recognized: one is based on the use of group-specific reagents, and the second utilizes affinity labels.

The modifiers belonging to the first class must be sufficiently reactive to bind covalently with an amino acid side-chain, and at the same time be sufficiently specific to react with only one type of side-chain. Many reagents for the modification of different amino acid side-chains have been developed and utilized; a large battery of compounds is therefore available for group-specific chemical modification studies on any enzyme. The principal problem encountered with this type of reagent is lack of specificity. Besides, usually more than one residue of a certain class is present in an enzyme molecule, making it difficult to achieve selectivity towards one single residue.

To overcome these limitations, researchers doing chemical modification increasingly turn to affinity labels. These compounds are chemically reactive analogs of enzyme ligands. Owing to this structural similarity, they show affinity to the ligand binding site, and therefore bind selectively on the enzyme surface. By means of their reactive group, these analogs can form a covalent bond with the enzyme. Since the specificity of these reagents is given by their affinity for a binding site, it is less important, and may even be desirable for them not to react specifically with only a certain type of amino acid side-chain. These reagents show saturation kinetics, and compete with their natural ligand counterpart for the binding site on the enzyme. The only important limitation of these reagents is that they must be prepared ad hoc for each enzyme or group of enzymes.

Two important subsets of affinity labels are the photoaffinity labels and the suicide inhibitors. Photoaffinity labels are activated by irradiation with light, and thus their reactivity can be closely controlled. The active species are free radicals which readily react with almost any chemical group in their neighborhood. These labels may also combine with non-polar side-chains. Suicide inhibitors are non-reactive compounds which act as substrates. By virtue of their transformation into products, a reactive group is generated, so that they readily bind to side-chains located in their vicinity at the active site. By their very nature, suicide inhibitors can be powerful pharmacological agents in the treatment of certain diseases.

The properties of group-specific reagents are discussed in more detail below. Affinity labels and photoaffinity labels are the subject of Chapters 3 and 4 in this volume. The reader interested in suicide inhibitors is referred to the review by Walsh (1984).

PROBLEMS AND LIMITATIONS IN CHEMICAL MODIFICATION

The investigator doing chemical modification should be aware of several limitations inherent in this technique. They have been aptly summarized by Cohen (1970), and are reviewed in the following paragraphs.

Few, if any, chemical modifiers are absolutely specific for a certain amino acid side-chain. This specificity can be influenced by the experimental conditions used, principally by the pH.

Absolute selectivity for one single amino acid residue is very seldom achieved. One particular enzyme may possess many residues of a certain kind, which can potentially react with a group-specific reagent. The reactivity of an amino acid side-chain is greatly influenced, however, by its microenvironment. At the active site of enzymes, which is in a less polar environment, amino acid residues often show a markedly different pKa than that of the same free amino acid. A good example is found in the enzyme oxaloacetate decarboxylase, possessing a lysine residue of pKa=5.9 (Schmidt & Westheimer 1971). These amino acids are often much more reactive, thus increasing the selectivity with which they may be labeled by group-specific reagents.

It is unlikely that any chemical modification can be accomplished without some change in the conformation of the protein. If conformational changes occur, they may be responsible for any observed change in the biological property being considered. Few investigators actually monitor conformational changes, although only minor changes have been observed in some cases studied (Kirschner & Schachman 1973, Horiike *et al.* 1979). The use of small reagents may help reduce conformational changes.

The behavior of a reagent towards a free amino acid is only a partial indication on how this reagent will react with the side-chain of the same amino acid in a protein, and the behavior towards one protein is only a partial indication of its behavior towards others. As indicated above, the microenvironment around a specific residue is a key factor in determining its reactivity. Several factors influence this microenvironment. One is polarity, which affects the pKa of the dissociable side-chains. Other important factors are hydrogen bonding effects, which may stabilize a neutral or ionic species, electrostatic effects (presence of charges in the vicinity of the group under study), and steric effects by other side-chains, which may hinder the approach of the reagent (Cohen 1970).

All the experimental variables can have an effect on the result of the modification. Temperature may be important if one is following the kinetics of modification. pH is another essential variable, since it influences the degree of ionization of both the reagent and the group under study. The nature of the buffer used is at times important. For example, more effective modification of arginines is obtained in borate than other buffers when butanedione is used as reagent, since borate stabilizes the reaction products (Riordan 1979).

TYPES OF INFORMATION OBTAINABLE FROM CHEMICAL MODIFICATION STUDIES

Stoichiometry of modification

It is very important to determine the number of molecules of modifier incorporated per active unit of enzyme in order to establish if non-specific modification has occurred. There are several means of accomplishing this, the choice depending on the type of reagent used. On occasions the formation of the covalent bond between modifier and target residue produces a change in the absorbtion spectrum of the modifier. This usually represents a simple way of quantifying the modification, although a knowledge of the extinction coefficient of the complex is essential; this value, however, may vary from enzyme to enzyme, as shown for the modification of lysine residues by pyridoxal phosphate (Bazaes *et al.* 1980).

Amino acid analysis of the modified enzyme can give clues as to the type and number of residues modified; the precision of this technique is of the order of 3–5%, so an accurate measurement of the disappearance of a residue may be difficult if the protein under study has a large content of these residues (Rohrbach & Bodley 1977). The appearance of a new product, such as S-carboxymethylcysteine in the modification of cysteine residues by iodoacetate, may be more easily detected.

The most common means of stoichiometry determination is to follow the incorporation of a radiolabeled modifier. It is more difficult, however, with this technique to determine the degree of non-specific modification.

Correlation of the degree of modification with change in a biological property (most commonly, enzyme activity)

A correlation of modification with alteration of a biological property is essential if one wants to ascribe a function to the modified residue. The simplest and ideal case is a linear relationship between these two variables. Usually such a correlation (or lack thereof) is established by means of a plot of residual activity against number of residues modified. The data can be extrapolated to zero residual activity to obtain the number of essential residues. If a linear response is obtained, one expects the modification to be an 'all or none' process, so that the total population of enzyme molecules is composed of a set of native and another of totally inactive molecules. This can be confirmed by determining the kinetic constants (Km and k_{cat}) of a partially modified sample of the enzyme. Horiike & McCormick (1979) have analyzed the validity of this approach and have pointed out that the extrapolation is valid if any non-essential residue reacts at least 100 times slower than the essential one.

If the above plot gives a non-linear result, the number of essential residues can still be established by means of the statistical analysis of Tsou (1962). This method is based on a plot of the total number of modified residues against the residual activity raised to the reciprocal power of a small integer ($i=1, 2, 3$, etc.). Several plots are made of the experimental data, each with a different value for i, and the resulting curves are analyzed statistically for the best fit to a straight line. The value of i corresponding to the best fit is the number of essential residues. Non-essential residues must react either much faster or much slower for the method to be valid. The reader is referred to Paterson & Knowles (1972) and Lundblad & Noyes (1984) for detailed description and applications of this interesting method.

Another method used to detect the identity and number of essential amino acid residues is the kinetic approach of Ray & Koshland (1961). This method is based on obtaining a relationship between the rate of loss of biological activity and the rate of modification of amino acid residues. To establish this relationship, the pseudo first-order rate constants of these processes must be determined. Limitations of this approach include the requirement of pseudo first-order kinetics (not always met in chemical modification), and the use of experimental conditions such that the velocity of modification can be accurately followed to allow a precise estimation of the constants.

Kinetic analysis of chemical modification

Such an analysis can give information on the kinetic mechanism of the inactivation process. It can also be used to determine the dissociation constant of a ligand–enzyme complex and of the pKa's of the reactive groups. The second application can be particularly useful when more conventional methods such as equilibrium dialysis cannot be employed, especially if the affinity for the ligand is low ($K_d > 10^{-3}$ M). All these methods require, however, that the inactivation process follows pseudo first-order kinetics. For more detail, see Chapter 2, by Cardemil, in this volume.

Sequencing of a peptide containing the modified amino acid

Recent advances in protein sequencing methodology (Han *et al.* 1985), especially the introduction of the gas-phase automatic sequenator (Hunkapiller *et al.* 1983) have greatly improved sequencing sensitivity and speed. The labeled protein can be subjected to chemical or proteolytic cleavage followed by separation of the labeled peptide. HPLC has proved particularly useful in this respect (Hunkapiller *et al.* 1984). Sequencing of the peptide can give information on the location of the essential residue in the protein primary structure. An example is the labeling and sequencing of a peptide from bovine muscle pyruvate kinase containing an essential lysine residue (Johnson *et al.* 1979).

Assignment of possible functions to the modified amino acid residue

Essential amino acid residues in the enzyme active site can participate in substrate binding or in catalysis. It is very difficult to assign these specific functions to a residue on the basis of chemical modification alone. However, certain types of experiment can give valuable information in this regard. Among them are protection experiments, in which the effect of substrate or other ligands on the velocity and the degree of modification is studied. The enzyme fructose-6-phosphate 2 kinase — fructose 2,6 bisphosphatase has been modified by pyridoxal 5' phosphate. Protection of the kinase activity is observed when modifying in the presence of fructose 2,6 bisphosphate and to a lesser extent by ATP. Two lysines are modified in the unprotected enzyme, while protection drastically reduces the modification, strongly suggesting that these residues are important for enzyme activity (Kitajima *et al.* 1985).

Sometimes a combination of chemical modification kinetics and substrate-catalyzed kinetics can reveal particularly useful mechanistic information. For example, this approach has provided evidence that catalytic competence in cysteine proteinases requires some structural change in addition to the formation of a thiolate$^-$/imidazolium$^+$ ion pair (Willenbrock & Brocklehurst 1984).

Correlation of the reversal of the modification and recovery of the altered biological function is additional useful evidence to establish the possible function of a residue. Bazaes *et al.* (1980) have shown that the modification of the only cysteine residue in liver phosphomevalonate kinase produces total loss in activity, and that activity is almost totally recovered if the modification is reversed by dithiothreitol, thus supporting a possible essential role for this residue at the active site.

To confirm the function assigned to a modified residue, it is very important to determine if the alteration caused by the modification is due to a steric effect. A very good approach to this question is to transform the putative essential residue into another amino acid of similar size and general properties. This can be accomplished (although not easily) by 'chemical mutation'. An example of such a study is the conversion of an essential cysteine of papain into serine, which resulted in a total loss of enzyme activity (Clark & Lowe 1978). Site-specific mutagenesis, produced by an 'in vitro' replacement of a purine or pyrimidine base in the protein's gene sequence, is a powerful technique to achieve these amino acid replacements; it has been used very successfully to study the function of amino acid residues in tyrosyl tRNA synthetase (Fersht *et al.* 1984) and other enzymes (Ackers & Smith 1985).

Study of allosteric and cooperative properties
Although chemical modification of enzymes is often associated with alterations at the enzyme active site, it has also been used successfully to analyze essential residues at allosteric sites. Kemp discusses this point in Chapter 7 of this volume. Residues essential for cooperative interactions have also been assigned by chemical modification, as shown by Slebe *et al.* (1983) for fructose 1,6 bisphosphatase. These authors show that loss in cooperativity by selective carbamylation of the AMP binding sites of this enzyme can be correlated to the modification of a lysine residue in each subunit.

Location of amino acid residues in the tertiary structure of a protein
The concepts of 'exposed' and 'buried' residues have been proposed to account for the degree of accessibility of residues to chemical modification. The 'exposed' residues are assumed to be located on the surface of the protein, while the 'buried' ones are hidden from the reagent in its hydrophobic interior. The 'buried' residues should become accessible as a result of denaturation of the enzyme. This topic is discussed further in Chapter 5, by Silva, in this volume.

Crosslinking by means of chemical modification
Reagents that contain two reactive functional groups can be used to form intramolecular or intermolecular crosslinks between amino acid residues of proteins. This approach is useful in establishing distances between the reactive side-chains, and can thus give insight into the tertiary or quaternary structure of a protein. A variety of reagents have been utilized for this purpose, among them 1,3 dibromoacetone, glutaraldehyde, dimethyl suberimidate, etc. A detailed account of these reagents and a survey of their application to the study of many proteins can be found in Lunblad & Noyes (1984). Photoaffinity crosslinking is discussed by Schaefer in Chapter 4 of this volume.

Alteration of enzyme specificity
Chemical modification of enzymes can change not only their catalytic parameters, but also their catalytic behavior in a qualitative manner (Kaiser *et al.* 1985). An interesting example is that of the flavopapains, where the

hydrolytic activity of papain is transformed into the oxido–reductase activity of a flavoprotein by alkylation of an active-site cysteine residue with a flavine derivative (Kaiser & Lawrence 1984).

SOME IMPORTANT REAGENTS USED IN THE MODIFICATION OF INDIVIDUAL AMINO ACID RESIDUES

As indicated above, polar amino acid residues which can show nucleophilicity at a certain pH range, can be subjected to chemical modification by group-specific reagents. These include the acidic amino acids glutamate and aspartrate, the basic amino acids lysine, arginine and histidine, the polar uncharged amino acids serine, cysteine and tyrosine, and the side-chains of methionine (possessing a nucleophilic sulfur) and tryptophan (heterocyclic indole side-chain). Numerous reagents have been introduced for the modification of these residues. A detailed review of them is beyond the scope of this chapter, so only a selected group is discussed in some detail. A good general review is found in Glazer (1976), Lundblad & Noyes (1984) give a comprehensive and up to date account of group-specific reagents, and Glazer *et al.* (1975) discuss experimental aspects in the use of many of these reagents.

Acidic amino acids (Glu and Asp)

The carboxylic side-chain of these amino acids has been modified by a number of different reagents. Hoare & Koshland (1966) introduced the use of water-soluble carbodiimides as catalysts for the specific modification of carboxyl groups in proteins by amines. This is the most widely used method. The binding of amines such as [^{14}C] glycine ethyl ester can be detected by the appearance of a new peak in amino acid analysis. Some reactivity by serine, cysteine, and tyrosine under certain experimental conditions has been reported (Carraway & Koshland 1972).

The conversion of carboxyl groups to esters by means of trialkyl oxonium fluoroborate salts has been used by several authors, among them Paterson & Knowles (1972). These investigators modified pepsin with [^{14}C] trimethyloxonium fluoroborate, establishing that at least two carboxyl groups are essential for the activity of the enzyme. Some reactivity is observed with methionine and histidine (Yonemitsu *et al.* 1969).

Isoxasolium salts were introduced by Woodward for the activation of carboxyl groups, initially for synthetic purposes (Woodward *et al.* 1961). Several of these derivatives have been used for chemical modification, especially N-ethyl-5-phenylisoxazolium-3'-sulfonate (Woodward's reagent K). Arana & Vallejos (1981) have used this reagent to study essential carboxyl groups in chloroplast coupling factor, comparing its effect with that of a carbodiimide. A method to quantify the degree of modification by spectrophotometric means has been described (Sinha & Brewer 1985).

Arginine

This amino acid possesses a highly basic guanidino group and has been proposed to play an esential role in the active site of enzymes binding anionic substrates (Riordan 1979). Most of the chemical modification studies of arginine residues have been done with dicarbonyl reagents, two of which, butanedione and phenylglyoxal, will be discussed further.

Butanedione was introduced by Yankeelov (Yankeelov *et al*. 1966), and made popular as an arginine modifier by Riordan, who observed that borate buffer stabilizes the arginine-butanedione complex (Riordan 1979), although this reaction is reversible. It has been pointed out that the reaction must be performed in the dark, since butanedione may act as a photosensitizing agent and cause the destruction of other residues, particularly tryptophan, histidine and tyrosine (Fliss & Viswanatha 1979, Gripon & Hofmann 1981). One of the many enzymes studied with this reagent is muscle pyruvate kinase (Cardemil & Eyzaguirre 1979): this enzyme is reversibly inactivated with concomitant loss in arginine as determined by amino acid analysis; protection experiments suggest that an essential arginine residue is located near the phosphate binding site of the substrate phosphoenolpyruvate.

Phenylglyoxal was first used by Takahashi (1968) as an arginine reagent. Two molecules of the reagent normally bind per arginine residue, and the reaction is irreversible. Reactivity with alfa-amino groups of peptides has been described (Takahashi 1968). Since [^{14}C]-phenylglyoxal can be easily synthesized (Schloss *et al*. 1978), its incorporation into proteins can readily be followed. Cheung & Fonda (1979) have shown that the reaction of phenylglyoxal with arginine in model compounds is faster in bicarbonate buffer, while α-amino groups appear unreactive under the same conditions.

Lysine

The epsilon amino group of this amino acid in its unprotonated form is a very reactive nucleophile in proteins. The pKa of this group is usually around 10. However, Lys residues of lower pKa can be found in enzymes as a result of microenvironmental effects; they are considerably more reactive and can therefore be selectively modified. A value of 5.9 has been found for a reactive lysine residue in oxaloacetate decarboxylase (Schmidt & Westheimer 1971).

Many compounds have been utilized for lysine modification. A useful reagent is trinitrobenzenesulfonate (TNBS). The reaction with amino groups causes specific spectral changes which can be followed at 420 or 367 nm (Lundblad & Noyes 1984). A highly reactive lysine residue of bovine muscle pyruvate kinase has been labeled with TNBS and a tryptic peptide containing the modified residue has been isolated and sequenced (Johnson *et al*. 1979).

Carbamylation of amino groups by means of cyanate has been used extensively. One clear advantage of this reagent is the small size of the cyanate ion. Reaction is also observed with histidine and cysteine, forming unstable carbamyl derivatives; carbamylation of the active-site serine of chymotrypsin (with concomitant loss of activity) has been reported (Shaw *et al.* 1964). Practical uses of this reagent are discussed by Stark (1972).

Pyridoxal phosphate (PLP), a natural derivative of vitamin B$_6$, is a very specific modifier of amino groups. The reaction leads to the reversible formation of a Schiff's base which can be stabilized by reduction with sodium borohydride. The stoichiometry of incorporation can be followed either spectrophotometrically (the reduced pyridoxyl derivative absorbs with a maximum at 325 nm (Glazer *et al.* 1975, page 131) or by reduction with ^{3}H-borohydride. A typical example of the use of this reagent in the determination of essential lysine residues is the work of Schnackerz & Noltmann (1971) on phosphoglucose isomerase. The authors find a very specific reaction with the incorporation of one molecule of PLP per subunit of enzyme.

Histidine
The role of histidine in enzyme-active sites has been reviewed by Schneider (1978). Two main procedures are used for the modification of the imidazole ring of histidine. Photo-oxidation was the first method introduced. Unfortunately, photo-oxidation shows low specificity, since methionine, tryptophan and to a lesser extent tyrosine, serine and threonine are also modified (Lundblad & Noyes 1984). Methylene blue and Rose Bengal are two of the commonly used dyes in this procedure.

Diethylpyrocarbonate is the most commonly used reagent for histidine modification (Miles 1977). This reagent shows good specificity near-neutral pH. The reaction leads to an increase in absorbance at 240 nm, and results in the substitution (carboxyethylation) of one of the imidazole nitrogens. The substitution can be reversed at alkaline pH, resulting in recovery of histidine. The reagent is somewhat unstable in aqueous media, especially at higher pH. Horiike *et al.* (1979) have used this reagent with pyridoxamine (pyridoxine)-5'-phosphate oxidase, and have shown that the enzyme is inactivated with the modification of a crucial His residue, without noticeable conformational perturbations. Dominici *et al.* (1985) have shown that diethylpyrocarbonate modifies one His residue of 3,4 dihidroxyphenylalanine decarboxylase which is essential for activity.

Cysteine
Owing to the high nucleophilicity of the thiol group (especially the thiolate anion), cysteine is the most reactive amino acid residue in proteins, and therefore can be modified by a great number of reagents. Detailed reviews of the reactions of thiol groups in proteins include those by Liu (1977) and Brocklehurst (1979).

Alkylating agents represent one of the most important groups of compounds used in thiol modification, especially iodoacetate and iodoacetamide. Iodoacetate is customarily used to carboxymethylate sulfhydryl groups of proteins prior to amino acid analysis or sequencing (Crestfield *et al.* 1963); this prevents Cys degradation and forms carboxymethyl cysteine which can be easily identified by the amino acid analyzer. Other haloacids or amides such as bromoacetic acid can be used, but the reaction is slower than with the iodo derivatives, because iodide is a better leaving group. Imidazole groups, although considerably less reactive, may also combine with haloacetic acids; an example of high imidazole reactivity is found in ribonuclease (Heinrikson *et al.* 1965).

N-ethylmaleimide has been a valuable reagent for sulfhydryl group modification. The reaction shows good specificity, and can be followed spectrophotometrically. The ethyl group can be replaced by a spin label, and this has been used effectively to study the mobility of the arm in the pyruvate dehydrogenase complex (Ambrose & Perham 1976).

Ellman (1959) introduced 5,5-dithiobis-(2-nitrobenzoic) acid (DTNB), which forms mixed disulfides when reacting with SH groups, liberating 2-nitro-5-thiobenzoic acid. This thiolate of the latter has a strong absorbance at 410 nm, and the reaction can therefore be easily followed spectrophotometrically. An example of the use of this reagent is the study of an essential cysteinyl residue in phosphomevalonate kinase (Bazaes *et al.* 1980). Applications of the more versatile disulfides containing the 2-mercaptopyridine group have been reviewed by Brocklehurst (1982). These disulfides act as 'two-protonic state' electrophiles, and this facilitates, among other applications, the isolation of thiol-enzymes by covalent chromatography (Brocklehurst *et al.* 1985).

Oxidation of thiol groups, other than by thiol-disulfide interchange, is another obvious chemical modification that would be expected to provide a high level of specificity. A particularly convenient chromophoric oxidizing agent for thiol groups is benzofuroxan (see Salih & Brocklehurst 1983).

Organic mercurials have been one of the oldest reagents used for cysteine modification, the most common one being *p*-chloromercuribenzoate. This compound becomes the hydroxy derivative when dissolved in water, and shows a strong increase in absorbance at 255 nm upon reacting with thiols. Bai & Hayashi (1979) have studied the effect of this compound on the activity of carboxypeptidase Y towards several substrates.

Tyrosine
Tetranitromethane has been a widely used reagent in tyrosine modification studies because of its high specificity and reactivity under mild conditions. This compound nitrates Tyr, producing an ionizable chromophore, 3-nitrotyrosine. Riordan *et al.* (1967) have explored the topology of the active site of nitrated carboxypeptidase A (which retains activity) following perturbations in the nitrotyrosyl spectrum through the binding of substrates and inhibitors.

Serine
This residue has been subjected to relatively few studies by means of group-specific chemical modification. A well-known case of participation of serine residues in enzyme active sites is in the so-called serine proteases, where a Ser residue shows high reactivity towards acylating reagents such as diisopropylfluorophosphate (Kraut 1977).

Methionine
This amino acid residue, although of low polarity, can be subjected to chemical modification owing to the nucleophilicity of the thioether sulfur. It is difficult to achieve selectivity when modifying this residue under mild conditions, but methionine can be oxidized to methionine sulfoxide by means of several reagents (Lundblad & Noyes 1984). Alkylation of this residue has also been reported, using for example, iodoacetate (Gundlach *et al.* 1959).

Tryptophan
The indole ring of tryptophan can be subjected to chemical modification. A commonly used reagent is *N*-bromosuccinimide (NBS), which oxidizes the indole residue to an oxindole derivative. The reaction can be monitored by following the decrease in absorbance at 280 nm. Reaction with tyrosine residues can occur, and this can interfere with the spectral measurements (Ohnishi *et al.* 1980). Warwick *et al.* (1972) have studied the effect of NBS on dihydrofolate reductase; they observed an initial increase in activity due to oxidation of a cysteinyl residue, followed at higher reagent concentrations by a loss in enzymatic activity associated with the oxidation of one Trp residue.

Koshland and co-workers have introduced the reagent 2-hydroxy-5-nitrobenzyl bromide for tryptophan modification (Horton & Koshland 1965). This reagent forms a hydroxynitrobenzyl derivative, which shows spectral properties that are sensitive to changes in the microenvironment (Lundblad & Noyes 1984).

CONCLUSION

This chapter is intended to serve as an overview of the methods used in the chemical modification of enzymes. The basic concepts, applications and limitations of chemical modification have been outlined. This technique, coupled with additional approaches can lead to an understanding of the structure and function of enzymes in solution.

REFERENCES

Ackers, G.K. & Smith, F.R. (1985) *Annual Review Biochem.* **54** 597–629.
Ambrose, M.C. & Perham, R.N. (1976) *Biochem. J.* **155** 429–432.
Arana, J.L. & Vallejos, R.H. (1981) *FEBS Letters* **123** 103–106.
Bai, Y. & Hayashi, R. (1979) *J. Biol. Chem.* **254** 8473–8479.
Bazaes, S., Beytia, E., Jabalquinto, A.M., Solis de Ovando, F., Gomez, I. & Eyzaguirre, J. (1980) *Biochemistry* **19** 2305–2310.
Brocklehurst, K. (1979) *Int. J. Biochem.* **10** 259–274.
Brocklehurst, K. (1982) *Methods Enzymol.* **87** 427–469.
Brocklehurst, K. Carlsson, J. & Kierstan, M.P.J. (1985) *Topics in Enzyme and Ferment. Biotech.* **10** 146–188.

Cardemil, E. & Eyzaguirre, J. (1979) *Arch. Biochem. Biophys.* **192** 533–538.

Carraway, K.L. & Koshland, D.E. Jr. (1972) *Methods Enzymol.* **25** 616–623.

Cheung, S.-T. & Fonda, M.L. (1979) *Biochem. Biophys. Res. Comm.* **90** 940–947.

Clark, P.I. & Lowe, G. (1978) *Eur. J. Biochem.* **84** 293–299.

Cohen, L.A. (1970) *The enzymes,* 3rd edn **1** 147–211.

Crestfield, A.M., Moore, S. & Stein, W.H. (1963) *J. Biol. Chem.* **238** 622–627.

Dominici, P., Tancini, B. & Voltattorni, C.B. (1985) *J. Biol. Chem.* **260** 10583–10589.

Ellman, G.L. (1959) *Arch. Biochem. Biophys.* **82** 70–77.

Fersht, A.R., Shi, J.P., Wilkinson, A.J., Blow, D.M., Carter, P., Waye, M.M.Y. & Winter, G.P. (1984) *Angewandte Chemie* (Int. edn) **23** 467–473.

Fliss, H. & Viswanatha, T. (1979) *Can. J. Biochem.* **57** 1267–1272.

Ford, L.O., Johnson, L.N., Mackin, P.A., Phillips, D.C. & Tjian, R. (1974) *J. Mol. Biol.* **88** 349–370.

Glazer, A. N. DeLange, R.J. & Sigman, D.S. (1975) *Chemical modification of proteins,* North-Holland/Elsevier, Amsterdam.

Glazer, A.N. (1976) *The proteins* 3rd edn **2** 1–103.

Gundlach, H.G., Moore, S. & Stein, W.H. (1959) *J. Biol. Chem.* **234** 1761–1764.

Gripon, J.-C. & Hofmann, T. (1981) *Biochem. J.* **193** 55–65.

Han, K.-K. Belaiche, D., Moreau, O. & Briand, G. (1985) *Int. J. Biochem.* **17** 429–445.

Heinrikson, R.L., Stein, W.H., Crestfield, A.M. & Moore, S. (1965) *J. Biol. Chem.* **240** 2921–2934.

Hoare, D.G. & Koshland, D.E. Jr. (1966) *J. Am. Chem. Soc.* **88** 2057–2058.

Horiike, K. & McCormick, D.B. (1979) *J. Theoret. Biol.* **79** 403–414.

Horiike, K., Tsuge, H. & McCormick, D.B. (1979) *J. Biol. Chem.* **254** 6638–6643.

Horton, H.R. & Koshland, D.E. Jr. (1965) *J. Am. Chem. Soc.* **87** 1126–1132.

Hunkapiller, M.V., Hewick, R.M., Dreyer, W.J. & Hood, L.E. (1983) *Methods Enzymol.* **91** 399–413.

Hunkapiller, M.V., Strickler, J.E. & Wilson, K.J. (1984) *Science* **226** 304–311.

Johnson, S.C., Bailey, T., Becker, R.R. & Cardenas, J.M. (1979) *Biochem. Biophys. Res. Comm.* **90** 525–530.

Kaiser, E.T. & Lawrence, D.C. (1984) *Science* **226** 505–511.

Kaiser, E.T., Lawrence, D.S. & Rokita, S.E. (1985) *Annual Review Biochem.* **54** 565–595.

Kirschner, M.W. & Schachman, H.K. (1973) *Biochemistry* **12** 2987–2996.

Kitajima, S., Thomas, H. & Uyeda, K. (1985) *J. Biol. Chem.* **260** 13995–14002.

Kraut, J. (1977) *Annual Review Biochem.* **46** 331–358.

Liu, T.-Y. (1977) *The proteins,* 3rd edn **3** 240–402.

Lundblad, R.L. & Noyes, C.M. (1984) *Chemical reagents for protein modification,* Vols 1 and 2, CRC Press, Boca Raton, Florida.

Miles, E.W. (1977) *Methods Enzymol.* **47** 431–442.

Ohnishi, M., Kawagishi, T., Abe, T. & Hiromi, K. (1980) *J. Biochem.* **87** 273–279.

Paterson, A.K. & Knowles, J.R. (1972) *Eur. J. Biochem.* **31** 510–517.

Quiocho, F.A. & Lipscomb, W.N. (1971) *Adv. Prot. Chem.* **25** 1–78.

Ray, W.J. Jr. & Koshland, D.E. Jr. (1961) *J. Biol. Chem.* **236** 1973–1979.

Riordan, J.F., Sokolovsky, M. & Vallee, B.L. (1967) *Biochemistry* **6** 358–361.

Riordan, J.F. (1979) *Mol. Cell. Biochem.* **26** 71–92.

Rohrbach, M.S. & Bodley, J.W. (1977) *Biochemistry* **16** 1360–1363.

Salih, E. & Brocklehurst, K. (1983) *Biochem. J.* **213** 713–718.

Schloss, J.V., Norton, I.L., Stringer, C.D. & Hartman, F.C. (1978) *Biochemistry* **17** 5626–5631.

Schmidt, D.E. & Westheimer, F.H. (1971) *Biochemistry* **10** 1249–1253.

Schnackerz, K.D. & Noltmann, E.A. (1971) *Biochemistry* **10** 4837–4843.

Schneider, F. (1978) *Angewandte Chemie* (Int. edn) **17** 583–592.

Shaw, D.C., Stein, W.H. & Moore, S. (1964) *J. Biol. Chem.* **239** PC671–673.

Sinha, U. & Brewer, J.M. (1985) *Anal. Biochem.* **151** 327–333.

Slebe, J.C., Herrera, R., Hubert, E., Ojeda, A. & Maccioni, R.O. (1983) *J. Prot. Chem.* **2** 437–443.

Stark, G.R. (1972) *Methods Enzymol.* **25** 579–584.

Takahashi, K. (1968) *J. Biol. Chem.* **243** 6171–6179.

Tsou, C.-L. (1962) *Sci. Sinica* **11** 1535–1558.

Walsh, C.T. (1984) *Annual Review Biochem.* **53** 493–535.

Warwick, P.E., D'Souza, L. & Freisheim, J.H. (1972) *Biochemistry* **11** 3775–3779.

Willenbrock, F. & Brocklehurst, K. (1984) *Biochem. J.* **222** 805–814.

Woodward, R.B., Olofson, R.A. & Mayer, H. (1961) *J. Am. Chem. Soc.* **83** 1010–1012.

Yankeelov, J.A., Jr., Kochert, M., Page, J. & Westphal, A. (1966) *Fed. Proc.* **25** 590.

Yonemitsu, O., Hamada, T. & Kanaoka, Y. (1969) *Tetrahedron Letters* 1819–1820.

2

Kinetics of the chemical modification of enzymes

Dr Emilio Cardemil, Departamento de Química, Universidad de Santiago de Chile, Santiago, Chile

INTRODUCTION

The purpose of this chapter is to summarize some methods currently employed for the analysis of the enzyme chemical modification kinetics, including the use of chemical modification as a tool to obtain information on quantitative aspects of enzyme–ligand interaction. Recent reviews on these subjects include those of Brocklehurst (1982), Plapp (1982) and Rakitzis (1984).

KINETIC MECHANISMS OF THE CHEMICAL MODIFICATION OF ENZYMES

The analysis of the kinetic mechanism of an enzyme chemical modification reaction depends on the conditions of the reaction system. The mathematical treatment is greatly simplified if pseudo-first-order conditions with respect to the modifier are used (that is, if the concentration of the modifier is much higher than the concentration of the enzyme).

Irreversible reactions

If we consider the simplest possible reaction for an enzyme (E) and an inactivator (I) to produce an inactive enzyme-inactivator complex (EI)

$$E + I \overset{k_1}{\to} EI \, ,$$

(1)

the reaction rate (v) is given by

$$v = k_1 \, [E][I] \, ,$$

(2)

and if $[I] \gg [E]$ it simplifies to

$$v = k_{obs}[E] \tag{3}$$

where

$$k_{obs} = k_1[I] \tag{4}$$

Equation (3) can also be written as

$$-\frac{d[E_a]}{dt} = k_{obs}[E_a] \tag{5}$$

where $[E_a]$ is the enzyme concentration at time t.

Integration of equation (4) with respect to $[E_a]$ between times 0 and t gives a linear relationship between the natural logarithm of the fractional activity of the enzyme and the time:

$$\text{Ln}\frac{[E_a]_t}{[E_a]_0} = -k_{obs}t, \tag{6}$$

so that a plot of the common logarithm of the remaining enzyme activity against time gives a straight line with slope $k_{obs}/2.303$. It is possible, then, to obtain k_1 from a series of determinations of k_{obs} at several different concentrations of I, according to equation (4). Deviations of linearity are expected, however, if the kinetic order of the reaction is not 1, as for example

$$E + 2I \rightarrow EI_2 \tag{7}$$

where

$$k_{obs} = k_1[I]^2 \tag{8}$$

In this particular situation a straight line can only be obtained by making a plot of k_{obs} against $[I]^2$ (Peters et al. 1981).

The kinetic order of a modification reaction is also commonly obtained from the logarithmic form of equation (8):

$$\log k_{\text{obs}} = n \log[\text{I}] + \log k_1 \qquad (9)$$

where n = kinetic order of the reaction or minimal number of I molecules needed to inactivate a single molecule of active enzyme unit. This relationship has been widely used since first employed by Levy *et al.* (1963), although not always properly, as pointed out by Jabalquinto *et al.* (1983) and by Carlson (1984). It must be pointed out that this relationship holds only for irreversible mechanisms such as that of equation (1), but not for reversible mechanisms or for mechanisms involving the formation of significant amounts of a dissociable complex prior to inactivation as those of equations (10) and (14). These mechanisms can be identified by appropriate treatment of the data, as will be seen below.

The second-order rate constant k_1 can also be obtained from experiments carried out in conditions where $[\text{E}] \approx [\text{I}]$ (as is done, for example, by Tian *et al.* (1985) for the inactivation reaction of chicken liver fatty acid synthetase by 5,5'-dithiobis-(2-nitrobenzoic acid), but in this case k_1 must be calculated from the slope of the appropriate second-order plot.

Reversible reactions
Let us now consider the mechanism

$$\text{E} + \text{I} \underset{k_{-1}}{\overset{k_1}{\rightleftharpoons}} \text{EI} \qquad (10)$$

where the reversible formation of an inactive EI complex is described. Introducing the condition that $[\text{I}] \gg [\text{E}]$, then

$$\text{E} \underset{k_{-1}}{\overset{k_1[\text{I}]}{\rightleftharpoons}} \text{EI} \qquad (11)$$

Now, if one assumes that the concentration of EI at the beginning of the reaction is zero, the rate equation can be easily integrated and simplified by introducing the equilibrium condition (Frost & Pearson 1961) to obtain

$$\text{Ln}\, \frac{([\text{E}] - [\text{E}_0])}{([\text{E}_0] - [\text{E}_e])} = (k_1[\text{I}] + k_{-1})\, t \qquad (12)$$

where $[\text{E}_0]$ is the initial concentration of E, $[\text{E}]$ is the concentration at any time t, and $[\text{E}_e]$ is the concentration at equilibrium. As the approach to equilibrium is a first-order process, the observed rate constant from a plot of $\text{Ln}\dfrac{[\text{E}]}{[\text{E}_0]}$ versus t will be the sum of the rate constants for the forward and reverse reactions (Strickland & Massey 1973):

$$k_{obs} = k_1[I] + k_{-1} \tag{13}$$

Therefore, a plot of k_{obs} as a function of [I] should be linear with a slope of k_1 and an extrapolated ordinate intercept of k_{-1}.

Formation of an intermediary complex
This kind of mechanism is the one expected for an affinity label (Knowles 1972, Plapp 1982) where a significant amount of dissociable complex is formed between the inactivator and the enzyme prior to the formation of an inactive EI complex:

$$E + I \underset{k_{-1}}{\overset{k_1}{\rightleftharpoons}} E{\cdot}I \overset{k_2}{\rightarrow} EI \tag{14}$$

For this mechanism, the observed rate of inactivation is given by

$$v = k_2 [E{.}I] \tag{15}$$

and, assuming that the equilibrium condition $k_2 \ll k_{-1}$ holds (Brocklehurst 1979), it can be shown (Kitz & Wilson 1962) that

$$k_{obs} = \frac{k_1 k_2 [I]}{k_1[I] + k_{-1}} \tag{16}$$

and, in reciprocal form,

$$\frac{1}{k_{obs}} = \frac{1}{k_2} + \frac{K_{diss}}{k_2} \cdot \frac{1}{[I]} \tag{17}$$

where $K_{diss} = k_{-1}/k_1$. Consequently, it can be seen now that for the mechanisms described in (1), (10), and (14), the direct plot of k_{obs} as a function of [I] should give, respectively: for (1) a straight line passing through the origin, with slope $= k_1$; for (10) a straight line with slope $= k_1$ and an extrapolated intercept $= k_{-1}$, and for (14) a rectangular hyperbola. This last case is best represented using the reciprocal equation (17), where slope $= K_{diss}/k_2$ and intercept $= 1/k_2$.

Discrimination among these three mechanisms is possible, then, by plotting k_{obs} versus [I] as shown in Fig. 2.1. It should be emphasized, however, that a wide enough range of concentrations of I must be tested in order to clearly differentiate among these mechanisms. For example, if the mechanism of equation (14) applies, but $[I] \ll k_{-1}/k_1$, equation (16) reduces to

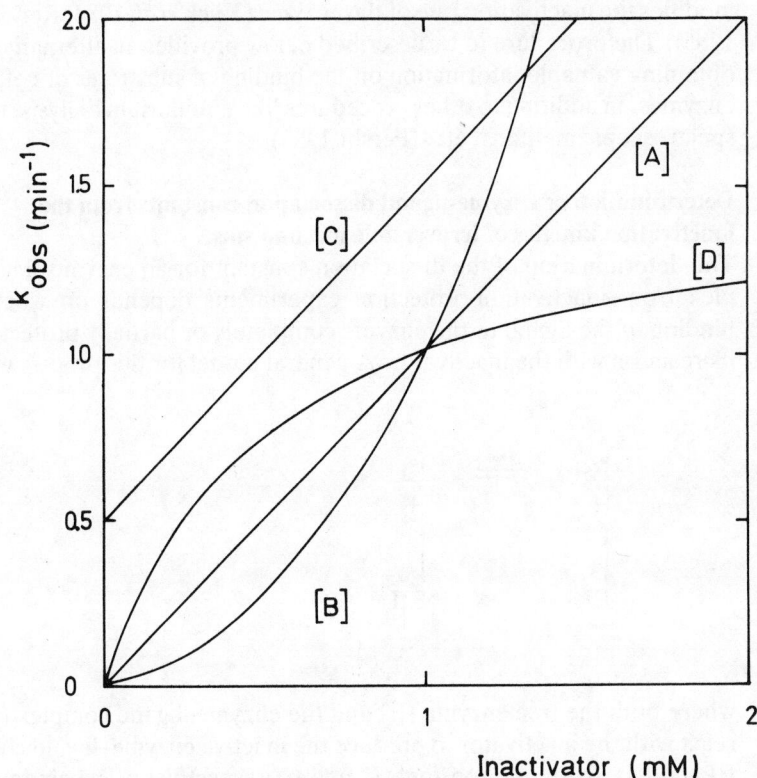

Fig. 2.1 — Plots of the variation of k_{obs} as a function of the concentration of inactivator. A, plot of equation (4); B, plot of equation (8); C, plot of equation (13); D, plot of equation (16). The values of the kinetic constants are: $k_1 = 1$ min^{-1} mM^{-1}; $k_{-1} = 1$ min^{-1}; $k_2 = 0.5$ min^{-1}.

$$k_{obs} = \frac{k_2}{K_{diss}} [I] \tag{18}$$

and a type A plot (Fig. 2.1) is obtained, even if an intermediate is involved in the inactivation reaction. Examples of enzyme inactivation through these mechanisms can be found in Jabalquinto *et al.* (1983), Cardemil & Eyzaguirre (1979) and Mas & Colman (1983).

USE OF KINETIC INACTIVATION TECHNIQUES FOR THE DETERMINATION OF ENZYME–LIGAND DISSOCIATION CONSTANTS.

Once the kinetic mechanism for the chemical modification of an enzyme by an inactivator is known, it is often possible to determine the dissociation constant for the binding of a ligand to the enzyme, provided the ligand modifies the inactivation rate of the enzyme (Kiick *et al.* 1984, Renosco *et al.* 1985). The procedure to be described below provides an alternative way of obtaining valuable information on the binding of substrates or cofactors to enzymes, in addition to other procedures like equilibrium dialysis, titration, spectroscopic methods, etc. (Fersht 1985).

Determination of enzyme-ligand dissociation constants from the inactivation kinetics of irreversible mechanisms.

The determination of the dissociation constant for an enzyme–ligand complex from inactivation-protection experiments depends on whether the binding of the ligand to the enzyme completely or partially protects it from its reaction with the inactivator. A general model for this process would be

$$
\begin{array}{ccc}
\text{E} + \text{L} & \xrightleftharpoons{K_{\text{diss}}} & \text{EL} \\
\big| + & & \big| + \\
\text{I} & & \text{I} \\
\big\downarrow{k_1} & & \big\downarrow{k_2} \\
\text{EI} & & \text{EIL}
\end{array}
\tag{19}
$$

where both the free enzyme (E) and the enzyme–ligand complex (EL) can react with the inactivator to produce the inactive enzyme–ligand complexes EI and EIL, with rate constants k_1 and k_2, respectively. The binding of L to E is dependent on the magnitude of the dissociation constant K_{diss}. If we assume that the equilibrium between E, L and EL is faster than the reaction of I with E, the inactivation rate of the enzyme in the presence of L can be expressed by

$$
- \frac{d[E_a]}{dt} = k_1[E][I] + k_2[EL][I]
\tag{20}
$$

where $[E_a]$ = concentration of active enzyme = $[E] + [EL]$.
   Dividing (20) by $[E_a]$ gives

$$
- \frac{d[E_a]}{dt[E_a]} = \frac{k_1[E][I] + k_2[EL][I]}{[E] + [EL]}
\tag{21}
$$

and as

$$
K_{\text{diss}} = \frac{[E][L]}{[EL]}
\tag{22}
$$

it follows that

$$- \frac{d[E_a]}{[E_a]} = \frac{[I](k_1 K_{diss} + k_2[L])}{K_{diss} + [L]} \, dt \tag{23}$$

So, the observed pseudo-first-order rate constant for the enzyme inactivation in the presence of a ligand (k_{obs}^L) is given by

$$k_{obs}^L = \frac{[I](K_{diss}k_1 + k_2[L])}{K_{diss} + [L]} \tag{24}$$

and, in reciprocal form,

$$\frac{1}{k_{obs}^L} = \frac{K_{diss}}{[I](K_{diss}k_1 + k_2[L])} + \frac{[L]}{[I](K_{diss}k_1 + k_2[L])} \tag{25}$$

or just

$$\frac{1}{k_{obs}^L} = \frac{1}{[I]k_1} + \frac{[L]}{[I]k_1 K_{diss}} \tag{26}$$

if $k_2 = 0$, when I does not react with EL.

Equations (25) and (26) predict a rectangular hyperbola and a straight line, respectively, when $1/k_{obs}^L$ is plotted against the concentration of L, thus providing a way to know if I reacts with the EI complex. When equation (26) applies, K_{diss} can be obtained from the extrapolated abscissas intercept of the plot (Mildvan & Leigh 1964). Fig. 2.2 shows experimental data for the determination of the dissociation constant of the yeast phosphoenolpyruvate carboxykinase-Mn^{2+} complex according to (26). When k_2 is not zero, K_{diss} is best estimated from the following relation introduced by Scrutton & Utter (1965):

$$\frac{k_{obs}^L}{k_{obs}} = \frac{k_2}{k_1} + \left[\frac{1 - \dfrac{k_{obs}^L}{k_{obs}}}{[L]} \right] K_{diss} \tag{27}$$

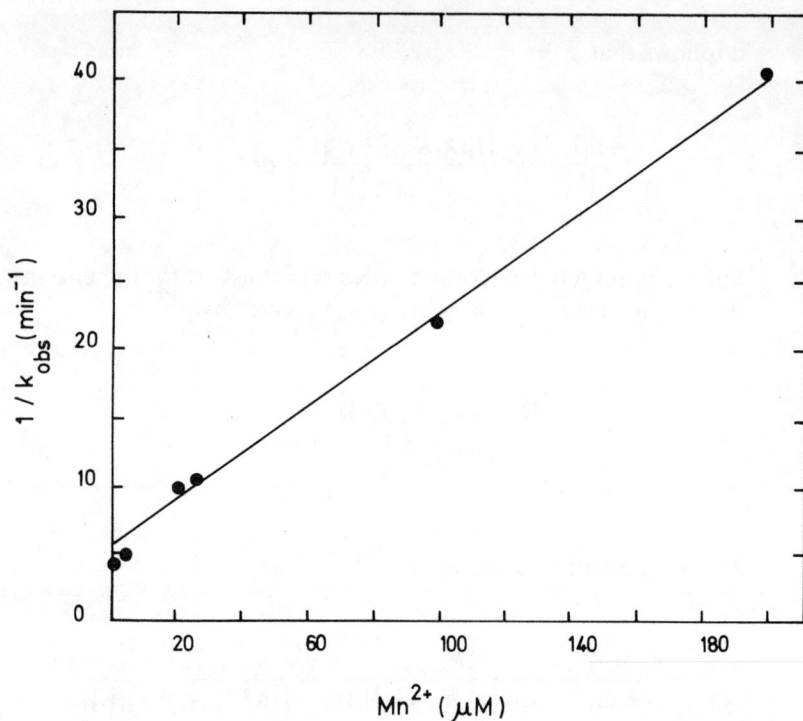

Fig. 2.2 — Dependence of $1/k_{obs}$ on the concentration of Mn^{2+} for the inactivation of yeast phosphoenolpyruvate carboxykinase by phenylglyoxal in 60 mM borate buffer pH 8.4 at 30°C. The least-square straight line fit of the data gives a dissociation constant of 28 μM, according to equation (26) (P. Malebrán and E. Cardemil, unpublished experiments).

Here, when k_{obs}^L/k_{obs} is plotted as a function of $\dfrac{1 - \dfrac{k_{obs}^L}{k_{obs}}}{[L]}$, K_{diss} can be obtained from the slope of the line. k_2 may be obtained from the intercept, since k_1 can be estimated from equation (4).

Another way of treating the data is provided by the relationship

$$k_{obs}^L = k_2[I] + \frac{k_{obs} - k_{obs}^L}{[L]} K_{diss} \tag{28}$$

employed, for instance, by Fujioka & Tanaka (1981) and Jabalquinto et al. (1983) for the determination of dissociation constants for enzyme-substrate complexes for yeast saccharopine dehydrogenase and chicken liver mevalonate-5-diphosphate decarboxylase, respectively.

Estimation of enzyme–ligand dissociation constants from the inactivation kinetics of reactions involving the formation of an intermediary complex.
This situation has been analyzed by Horiike & McCormick (1980) and by Carrillo *et al.* (1981). A general scheme for this inactivation mechanism in the presence of a ligand is

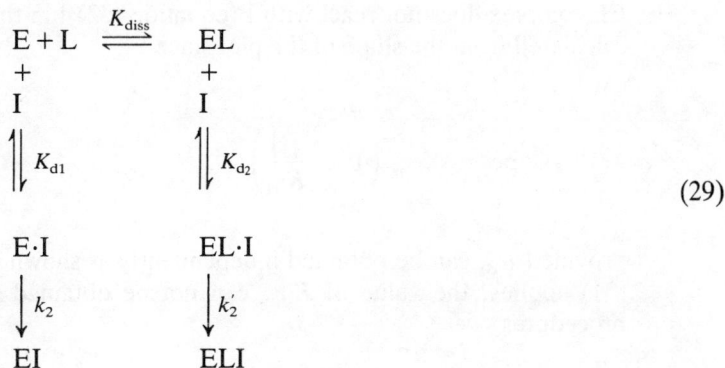

$$
\begin{array}{ccc}
\mathrm{E+L} & \underset{\longleftarrow}{\overset{K_{\mathrm{diss}}}{\rightleftharpoons}} & \mathrm{EL} \\
+ & & + \\
\mathrm{I} & & \mathrm{I} \\
\Big\downarrow K_{\mathrm{d1}} & & \Big\downarrow K_{\mathrm{d2}} \\
\mathrm{E \cdot I} & & \mathrm{EL \cdot I} \\
\Big\downarrow k_2 & & \Big\downarrow k_2' \\
\mathrm{EI} & & \mathrm{ELI}
\end{array}
\tag{29}
$$

where it is assumed again that both E and EL can react with I to produce inactive EI and ELI complexes through the active E.I and EL.I complexes, respectively. K_{d1} and K_{d2} are dissociation constants while k_2 and k_2' are rate constants.

A general rate equation for mechanism (29) has been derived by Carrillo *et al.* (1981) for the situation where both [I] and [L] are large compared to [E], and from that equation it can be deduced that

$$
k_{\mathrm{obs}}^{\mathrm{L}} = \frac{(k_2 K_{\mathrm{diss}} K_{\mathrm{d2}} + k_2'[\mathrm{L}] K_{\mathrm{d1}})[\mathrm{I}]}{K_{\mathrm{d2}} K_{\mathrm{diss}}(K_{\mathrm{d1}} + [\mathrm{I}]) + K_{\mathrm{d1}}[\mathrm{L}](K_{\mathrm{d2}} + [\mathrm{I}])}
\tag{30}
$$

or, in a reordered form,

$$
\frac{k_{\mathrm{obs}}^{\mathrm{L}}}{k_{\mathrm{obs}}} = \frac{k_2'(K_{\mathrm{d1}} + [\mathrm{I}])}{k_2(K_{\mathrm{d2}} + [\mathrm{I}])} + K_{\mathrm{d}}\frac{K_{\mathrm{d2}}(K_{\mathrm{d1}} + [\mathrm{I}])}{K_{\mathrm{d1}}(K_{\mathrm{d2}} + [\mathrm{I}])} \cdot \frac{\left(1 - \dfrac{k_{\mathrm{obs}}^{\mathrm{L}}}{k_{\mathrm{obs}}}\right)}{[\mathrm{L}]}
\tag{31}
$$

or just

$$
\frac{k_{\mathrm{obs}}^{\mathrm{L}}}{k_{\mathrm{obs}}} = K_{\mathrm{diss}}\frac{(K_{\mathrm{d1}} + [\mathrm{I}])}{K_{\mathrm{d1}}} \cdot \frac{\left(1 - \dfrac{k_{\mathrm{obs}}^{\mathrm{L}}}{k_{\mathrm{obs}}}\right)}{[\mathrm{L}]}
\tag{32}
$$

for the case where the EL complex does not react with I (that is, for the situation where $k_2' = 0$ and K_{d2} tends to infinity).

By comparing equations (31) and (32) it is clear that the extrapolation of a plot of k_{obs}^L/k_{obs} versus $(1 - k_{obs}^L/k_{obs})/[L]$ passes through the origin if the EL complex does not react with I (equation (32)); in this case K_{diss} can be calculated from the slope of the plot since

$$\text{slope} = K_{diss}\left(1 + \frac{[I]}{K_{d1}}\right) \tag{33}$$

provided K_{d1} can be obtained independently as shown in (16) and (17). If (31) applies, the value of K_{diss} can not be obtained by these graphical procedures.

EFFECT OF pH ON ENZYME CHEMICAL MODIFICATION REACTIONS

The determination of the rate of the chemical modification of an enzyme by an inactivator as a function of pH allows one to identify the pK of the reactive group in the enzyme.

The determination of pK values for specific amino acid residues in proteins by means of chemical modification is a valuable piece of information, since it allows one to verify pK values assigned to these residues by other methods. For example, it is often possible to obtain pK values for catalytically important groups in enzymes from studies of the pH effect on k_{cat}, K_m, or k_{cat}/K_m (Fersht 1985). A comparison, then, of the pK values obtained from both kinetic and chemical modification studies provides a good opportunity for corroborating specific aspects of the chemical mechanism of an enzyme. Using this approach, Willenbrock & Brocklehurst (1984) have postulated a detailed scheme for the protonic dissociation in cathepsin B (a cysteine proteinase) explaining thus the nucleophilic character and catalytic activity of the enzyme.

The mathematical treatment for most cases is identical to that described previously for the determination of K_{diss} for the binding of a ligand. We can thus obtain (from 26) the following relationship

$$\frac{1}{k} = \frac{1}{\tilde{k}} + \frac{[H^+]K_A}{\tilde{k}} \tag{34}$$

where k is the pseudo-first-order inactivation rate constant at a given H^+ concentration, $\tilde{k}$ is the pH-independent pseudo-first-order inactivation rate constants, and K_A is the equilibrium constant for the binding of H^+ to the enzyme. A plot of $1/k$ as a function of $[H^+]$ allows the determination of K_A from the extrapolated abscissas intercept of the graph (Tian *et al.* 1985).

If the inactivation of the enzyme follows a mechanism that includes the formation of a reversible complex between enzyme and inactivator like

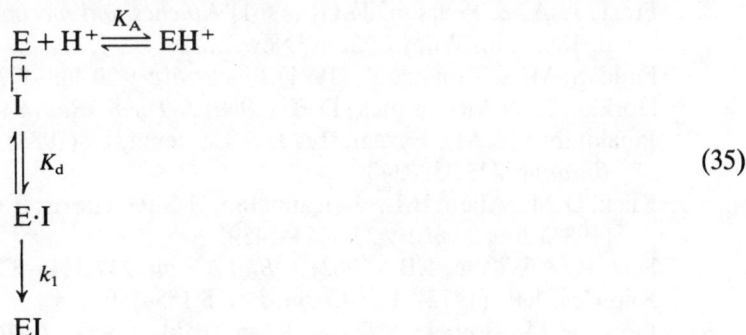

$$
\begin{array}{c}
\mathrm{E} + \mathrm{H}^+ \xrightleftharpoons{K_A} \mathrm{EH}^+ \\
\big|\!+ \\
\mathrm{I} \\[4pt]
\Big\Updownarrow K_\mathrm{d} \\[4pt]
\mathrm{E\cdot I} \\[4pt]
\Big\downarrow k_1 \\[4pt]
\mathrm{EI}
\end{array}
\tag{35}
$$

the following relation, obtained from (32), holds

$$
k = \frac{\dfrac{k_1[\mathrm{I}]}{K_A}}{\dfrac{K_\mathrm{d}+[\mathrm{I}]}{K_A} + K_\mathrm{d}[\mathrm{H}^+]}
\tag{36}
$$

or, in reciprocal form,

$$
\frac{1}{k} = \frac{1}{k_1} + \left(\frac{K_\mathrm{d}}{k_1} + \frac{K_\mathrm{d}K_A}{k_1}[\mathrm{H}^+] \right) \frac{1}{[\mathrm{I}]}
\tag{37}
$$

Thus, a plot af $1/k$ as a function of $1/[\mathrm{I}]$ at a series of concentrations of H^+ will yield a family of straight lines with a common intercept on the $1/k$ axis of $1/k_1$. Replots of the slopes of these lines against the concentration of H^+ will provide the value of K_A from the extrapolated intercept on the $[\mathrm{H}^+]$ axis.

Equations for more complex cases (i.e. the protonated form of the enzyme also binds the inactivator) can also be derived; the reader is referred to Tipton & Dixon (1979) for details.

REFERENCES

Brocklehurst, K. (1979) *Biochem. J.* **181** 775–778.

Brocklehurst, K. (1982) *Methods Enzymol.* **87** 427–469.

Cardemil, E. & Eyzaguirre, J. (1979) *Arch. Biochem. Biophys.* **192** 533–538.

Carlson, G.M. (1984) *Biochim. Biophys. Acta* **789** 347–350.

Carrillo, N., Arana, J.L. & Vallejos, R.H. (1981) *J. Biol. Chem.* **256** 6823–6828.

Fersht, A. (1985) *Enzyme structure and mechanism*, 2nd edn, W.H. Freeman, San Francisco.

Frost, A.A. & Pearson, R.G. (1961) *Kinetics and mechanisms*, 2nd edn, p. 186, John Wiley & Sons, New York.

Fujioka, M. & Tanaka, Y. (1981) *Biochemistry* **20** 468–472.

Horiike, K. & McCormick, D.B. (1980) *J. Theor. Biol.* **84** 691–708.

Jabalquinto, A.M., Eyzaguirre, J. & Cardemil, E. (1983) *Arch. Biochem. Biophys.* **225** 338–343.

Kiick, D.M., Allen, B.L., Jangannatha, G.S.R., Harris, B.G. & Cook, P.F. (1984) *Biochemistry* **23** 5454–5459.

Kitz, R. & Wilson, I.B. (1962) *J. Biol. Chem.* **237** 3245–3249.

Knowles, J.R. (1972) *Acc. Chem. Res.* **5** 155–160.

Levy, H.M., Leber, P.D. & Ryan, E.M. (1963) *J. Biol. Chem.* **238** 3654–3659.

Mas, M.T. & Colman, R.F. (1983) *J. Biol. Chem.* **258** 9332–9338.

Mildvan, A.S. & Leigh, R.A. (1964) *Biochim. Biophys. Acta* **89** 393–397.

Peters, R.G., Jones, W.C. & Cromartie, T.H. (1981) *Biochemistry* **20** 2564–2571.

Plapp, B.V. (1982) *Methods Enzymol.* **87** 469–499.

Rakitzis, E.T. (1984) *Biochem. J.* **217** 341–351.

Renosco, F., Seubert, P.A., Knudson, P. & Segel, I.H. (1985) *J. Biol. Chem.* **260** 11903–11913.

Scrutton, M.C. & Utter, M.F. (1965) *J. Biol. Chem.* **240** 3714–3723.

Strickland, S. & Massey, V.S. (1973) *J. Biol. Chem.* **248** 2953–2962.

Tian, W.X., Hsu, R.Y. & Wang, Y.S. (1985) *J. Biol. Chem.* **260** 11375-11387.

Tipton, K.F. & Dixon, H.B.F. (1979) *Methods Enzymol.* **63** 183–234.

Willenbrock, F. & Brocklehurst, K. (1984) *Biochem. J.* **222** 805–814.

3

Affinity labels as probes of the active site of enzymes. The use of dialdehyde derivatives of nucleotides

Dr Sergio Bazaes, Laboratorio de Bioquímica, Universidad Católica, Santiago, Chile

GENERAL CONCEPTS

Enzymes are highly specific and efficient biological catalysts. This is primarily due to the capacity of the enzyme active site to form specific complexes with its substrates to promote chemical reactivity.

Numerous studies have been performed on the active site of enzymes in order to understand their mechanisms of action. In regulatory enzymes the structure of sites that bind allosteric effectors is also important. One method of obtaining information about active or allosteric sites of enzymes is through chemical modification. This may be accomplished by using specific reagents directed to distinct chemical groups in the protein, or by means of affinity labels. The goal is to produce a change in some property of the enzyme that can be correlated with the functional role of specific amino acid residues. The specific chemical modification ideally results in the quantitative modification of a functional group belonging to a unique amino acid residue without affecting other functional groups or the conformation of the enzyme molecule. The limitations of the chemical modification approach in active-site studies is discussed in Chapter 1 of this work.

Affinity labels have proved to be powerful tools in the study of the relationship between structure and function of proteins. These reagents, also called site-specific reagents, are characterized by being structurally similar to known substrates, allosteric effectors or other ligands that bind to proteins, and therefore show affinity towards these ligands binding sites. These compounds are being extensively studied, and many affinity reagents have been designed for a number of proteins. Several reviews, some very recent ones, have appeared on the subject (Shaw 1970, Colman *et al.* 1977, Plapp 1982, Colman 1983), including a special volume of *Methods in Enzymology* (Jacoby & Wilchek 1977).

Affinity labels differ from reversible inhibitors in that they possess reactive groups capable of forming covalent bonds with amino acid side-chains. The reactive groups are usually alkylating or arylating agents. A very important class of affinity reagents are the photoaffinity labels, which have a photolabile group in their structure. These agents are activated by irradiation, forming 'in situ' a very reactive functional group (usually a carbene or nitrene) which is capable of interacting with the neighboring chemical residues. In this way it is possible to modify hydrophobic or polar residues which are normally of low reactivity (Glazer *et al.* 1975). Photoaffinity reagents are further discussed by Schafer in this work.

Affinity labeling involves at least two steps: the site-specific binding of the reagent and the subsequent modification of an amino acid residue through the formation of a covalent bond. Affinity labeling must show saturation kinetics (Meloche 1967, Wold 1977). It can be demonstrated that the kinetic process corresponds to the following scheme:

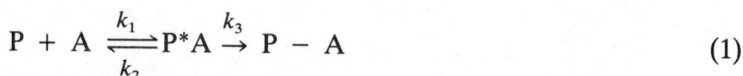

$$P + A \underset{k_2}{\overset{k_1}{\rightleftharpoons}} P^*A \overset{k_3}{\rightarrow} P - A \qquad (1)$$

$$K_a = \frac{[P][A]}{[P^*A]} = \frac{k_2}{k_1} \qquad k_{obs} = \frac{k_3[A]}{[A] + K_a} \qquad (2)$$

where P is the protein, A the affinity label, P^*A the non-covalent (Michaelis-type) complex, $P - A$ the covalent product of the reaction and K_a the dissociation constant of the non-covalent complex. The designation of K_a as k_2/k_1 depends on an assumption of quasi-equilibrium around P^*A. This is probably a valid assumption except for very rapid modification reactions where the apparent second-order rate constant under conditions of low [A] is greater than ca. $10^6 \, M^{-1} \, s^{-1}$ (Brocklehurst 1979).

The pseudo-first-order inactivation rate constant (k_{obs}) is obtained upon conditions where the modifying reagent is present in great excess in relation to the protein. According to equation (2), the pseudo-first-order inactivation rate constant presents saturation kinetics at high reagent concentration; this is one of the requirements that a reagent must satisfy in order to be considered as affinity label. By the very nature of the phenomenon under study, it is expected that a substrate, competitive inhibitor or ligand will protect against modification and inactivation. In addition, the reaction must be stoichiometric to the sites modified in the functional subunit.

USE OF PERIODATE-OXIDIZED NUCLEOTIDES TO LABEL PURINE-NUCLEOTIDE BINDING SITES IN PROTEIN.

Purine nucleotides are very important compounds in cellular metabolism. They are involved in the reaction of kinases, and they function as allosteric effectors of many enzymes. Moreover, as part of NAD or NADP, they participate in reactions catalyzed by dehydrogenases.

Many reagents have been designed and used to label the nucleotide binding site of enzymes, including alkyl halide derivatives of purine nucleotides (Bednar & Colman 1982, Roy & Colman 1980, Severin et al. 1976), fluorosulfonylbenzoyl analogs (Pal et al. 1975, Roy & Colman 1979, Annamalai & Colman 1981, Savadambal et al. 1981), photoaffinity purine nucleotide analogs such as 8-azido nucleotide analogs (Scheurich et al. 1978, Marcus & Haley 1979, Schaefer et al. 1980, Taylor & Kerlavage 1982), and periodate-oxidized nucleotides. This last group of compounds presents several advantages such as easy of preparation and solubility in aqueous solvents similar to that of the non-modified nucleotides. They can also be easily prepared as radioactive labels. In general, periodate-oxidized nucleotides behave as affinity labels of enzymes, although a few cases of non-specific reactions are reported (Mehler et al. 1981).

The oxidation of ribonucleotides by periodate results in cleavage between the 2'-3' carbons of the ribose moiety to yield the corresponding dialdehyde derivatives. This destruction of the ribose ring could be important in the binding of the nucleotide to certain enzymes. Aldehydes are capable of reacting with primary amines such as the ε-amino group of lysine or the terminal α-amino group of proteins. Periodate-oxidized ATP has been extensively characterized by NMR spectroscopy (Lowe & Beechey 1982a).

When studying the effect of 2',3' dialdehyde ATP (oATP) on sheep liver mitochondrial pyruvate carboxylase (Easterbrook-Smith et al. 1976) it was found that the Mg-oATP^{-2} complex behaved as a competitive reversible inhibitor with respect to MgATP^{-2}. When NaBH$_4$ was added to the reaction mixture, Mg-oATP became covalently bound to the enzyme, producing irreversible inactivation. These results agreed with the formation of a Schiff's base, which is reduced by NaBH$_4$. MgATP^{-2} protected the enzyme against chemical modification, suggesting that oATP was binding to the nucleotide site. Total loss of activity was observed when one mol of Mg-oATP was incorporated per mol of enzyme subunit. The modified amino acid was identified as lysine following enzymatic digestion of the Mg[^{14}C]oATP^{-2}-labeled enzyme and chromatography using [^{3}H]Lys-oATP as standard.

The use of periodate-oxidized NADP$^+$ (oNADP) was introduced by

Dallocchio *et al.* (1976). They showed that oNADP could be bound to *Candida utilis* 6-phosphogluconate dehydrogenase producing reversible inactivation of the enzyme. The inactivation was made irreversible by $NaBH_4$ reduction. These investigators chemically synthesized the two expected products from the reaction between the aldehyde groups of carbon 2' or 3' of the modified ribose and a lysine residue of the enzyme and its subsequent reduction with $NaBH_4$ and hydrolysys. Chromatographic analysis showed that the products derived from the reaction between the enzyme and oNADP behaved identically to the chemically synthesized standards. These derivatives have been separated and identified by amino acid analysis (King & Colman 1983).

Another communication reporting the formation of a Schiff's base between enzyme and oxidized nucleotides (Kumar *et al.* 1979) shows that oATP behaves as a good affinity label for the latent ATPase from *Mycobacterium phlei*. This reagent produced the progressive inactivation of latent and unmasked activities of the ATPase with non-linear dependence between the pseudo-first-order inactivation rate constant and oATP concentration. These results indicate the formation of a reversible complex prior to the covalent modification. The substrate ATP protected the enzyme against inactivation. At 100% inactivation, one mol of oATP was incorporated per mol of enzyme.

The effect of oATP on partially purified adenylate cyclase from bovine brain has also been studied (Wescott *et al.* 1980). The analog behaved as a competitive inhibitor of the enzyme. The simultaneous treatment with oATP and $NaCNBH_3$ produced irreversible inactivation of the enzyme with pseudo-first-order kinetics. A hyperbolic relationship between inactivation constants and oATP concentration was found. ATP and Tris protected the enzyme against inactivation. These data suggest the formation of a Schiff's base between the enzyme and oATP, although this could not be verified because of lack of pure enzyme.

Chemical bonds different from Schiff's bases have also been detected in the reaction between an enzyme and a periodate-oxidized nucleotide. An example is found in rabbit muscle phosphofructokinase (PFK) treated with dialdehyde-ATP (Gregory & Kaiser 1979). oATP produced enzyme inactivation which was not reversed by dialysis. A 99% inactivation was achieved with the simultaneous incorporation of 3–4 moles of the analog per subunit. ADP and ATP partially protected the enzyme from inactivation. Amino acid analysis of the modified enzyme did not show any difference from the untreated enzyme, whether or not $NaBH_4$ was added. These results suggest that the product of the reaction between oATP and PFK does not correspond to a Schiff's base. Furthermore, the modified form of PFK was stable in 0.01 N HCl, which is inconsistent with the lability expected for a Schiff's base. These authors proposed that the lysine residues of PFK react with oATP forming morpholine-type adducts, a result not previously observed between lysine residues of proteins and aldehyde reagents.

The 2'3'-dialdehyde derivative of ATP produced the inactivation of skeletal muscle phosphorylase kinase (King & Carlson 1981). The reaction was pseudo-first-order with saturation kinetics and the natural substrate, ATP, protected against inactivation. Furthermore, oATP could be used as a substrate to phosphorylate phosphorylase b, emphasizing the suggestion that oATP binds to the active site of the enzyme. The oATP-inactivated enzyme could not be reactivated by dilution with buffer or by dialysis. The treatment of the enzyme-oATP complex with $NaBH_4$ or $NaCNBH_3$ did not affect the degree of inactivation. These results suggest that oATP binds strongly to the enzyme, but the inactivation product is not a Schiff's base.

The 2',3'-dialdehyde derivative of 8-azido adenosine 5'-triphosphate (8-N_3oATP) has been used on phosphorylase kinase (King et al. 1982). This compound possesses the dialdehyde groups of oATP and is also a photoaffinity reagent containing an azido group. The enzyme became inactivated in the presence of the analog in the absence of photolysis, and showed similar behavior to the previously reported results with oATP (King & Carlson 1981). The results showed a similar K_i and the ability to serve as substrate. Upon irradiation the analog was incorporated to the enzyme, preferentially labeling the β subunit, similar to the effect observed with 8-azido adenosine 5'-triphosphate. This latter compound prevented the incorporation of 8-N_3oATP into the enzyme, suggesting that both azido analogs compete for the same binding site.

Pig heart NAD-dependent isocitrate dehydrogenase was modified by oADP (King & Colman 1983). The enzyme was inactivated with biphasic pseudo-first order kinetics. For both phases a non-linear relationship between rate constant and reagent concentration was found. ADP and isocitrate in the presence of Mn^{+2} produced a significant reduction in the rate of inactivation, and ADP seemed to compete with the analog for the same nucleotide binding site. The incorporation of approximately 1 mol of [^{14}C]oADP per mol average subunit produced the complete inactivation of the enzyme. The protease-digested [^{14}C]oADP-inactivated enzyme did not form the lysine derivatives expected for a Schiff's base between oADP and an amino group of the enzyme; all the radioactivity was associated with neutral amino acids even though a negatively charged product was expected. Moreover, the reduction of the oADP-labeled enzyme with NaB^3H_4 did not incorporate ^3H to the enzyme. These results provide evidence against a Schiff's base product for the inactivation of isocitrate dehydrogenase by oADP, and suggest a reaction mechanism in which both aldehyde groups become unavailable for reduction.

In a previous study on the inactivation of mitochondrial adenosine triphosphatase by oATP (Lowe et al. 1979) it was proposed that after the binding of oATP to the enzyme an elimination reaction occurs, which liberates the triphosphate group and forms a very stable conjugate Schiff's base. Therefore, isocitrate dehydrogenase was modified with both [^{14}C]oADP and [^{32}P]oADP under standard conditions to test for β elimination of the pyrophosphate group from oADP bound to the enzyme. The authors found a dramatic difference between binding of [^{14}C]oADP and [^{32}P]oADP per mol of average subunit, [^{32}P]oADP being bound in very small amounts. They also showed that the release of [^{32}P]-pyrophosphate occurred after oADP binding to the enzyme. The eliminaton reaction may lead to the formation of a stable conjugate Schiff's base which may not require reduction by $NaBH_4$ to stabilize the analog on the enzyme (Lowe & Beechey 1982b). This type of reaction should have allowed for incorporation of ^3H into the unreacted 2' aldehyde group, however, when the oADP-labeled isocitrate dehydrogenase was reduced with NaB^3H_4. A possible reaction product involving both aldehyde groups of oADP is a dihydroxymorpholine derivative, as suggested for phosphofructokinase inactivated by oATP (Gregory & Kaiser 1979). This derivative is relatively stable in dilute acid (0.01 N HCl), but the hydrolysis with 6 N HCl regenerates the unmodified lysines. The reaction product between oADP and isocitrate dehydrogenase most consistent with the available data is a 4',5'-didehydro-2',3'-dihydroxymorpholine derivative of lysine. In spite of the relative stability of the product, attempts to separate the proposed derivative have been unsuccessful.

In our laboratory rabbit muscle pyruvate kinase has been modified with oADP (Hinrichs & Eyzaguirre 1982). The enzyme is inactivated in a first-order reaction with saturation kinetics. These results indicate that the enzyme forms a reversible complex prior to covalent modification. ADP and ATP, especially in the presence of Mg^{+2}, protect the enzyme against inactivation. oADP is not a substrate, but acts as a competitive inhibitor with respect to ADP. The reaction between oADP and pyruvate kinase seems to be very specific. Fig. 3.1 shows that in the presence of 25 mM MgATP the incorporation of $[^{14}C]$oADP is very low; in the absence of the protective compounds the enzyme rapidly incorporates about 1 molecule of $[^{14}C]$oADP per subunit. The inactivation reaction is neither reversed by removal of excess oADP by exhaustive dialysis or gel filtration, nor is any reactivation observed after addition of Tris or lysine to the inactivation mixture. The addition of $NaBH_4$ was not necessary to stabilize the linkage formed. In fact, the addition of NaB^3H_4 to the oADP-modified pyruvate kinase does not incorporate ^3H to the enzyme. The mechanism for the reaction of oADP with pyruvate kinase is therefore most consistent with the formation of a morpholine derivative.

The peptide containing the radioactive analog was isolated and sequenced (Bezares *et al.* 1987). For that purpose the enzyme was incubated with non-radioactive oADP in the presence of MgATP. Under these conditions the enzyme is not inactivated. After removal of the protective effectors, the enzyme was incubated with $[^{14}C]$oADP, and now labeling and inactivation occurred. Fig. 3.2 shows that the incorporation of approximately 1 mol of the analog per subunit correlates with total loss of activity.

The modified enzyme was reduced, carboxymethylated, and digested with trypsin. The isolation of the oADP-peptide was performed by chromatography on Sephadex G-25, Sephadex G-50, and HPLC. The purified peptide was sequenced by the automatic gaseous-phase Edman method. The complete sequence of a 34 amino acid residue peptide was obtained, but no radioactive label was found (Bezares *et al.* 1987). The sequence obtained for the peptide was identical to a peptide previously isolated from bovine

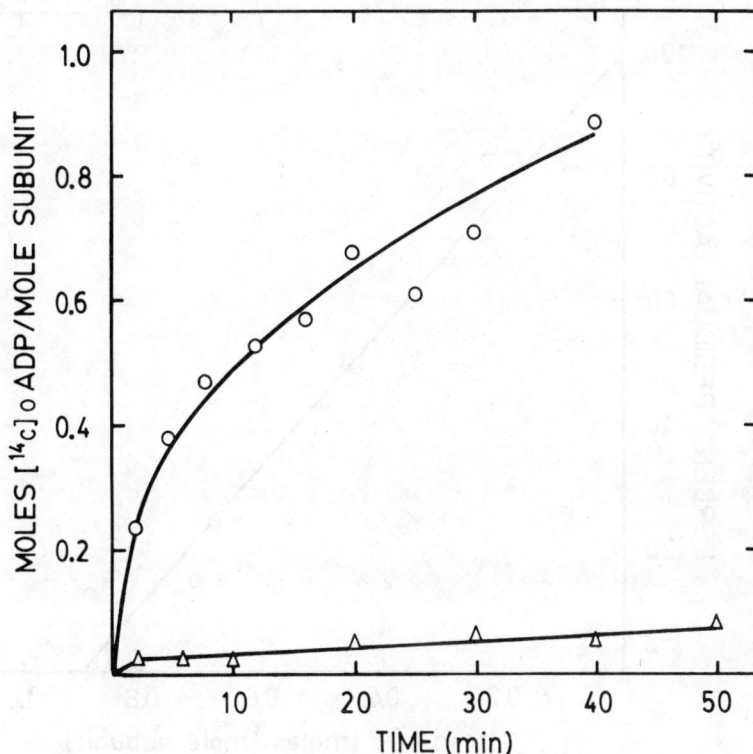

Fig. 3.1 — Time-course of the incorporation of [^{14}C]oADP into pyruvate kinase. Purified rabbit muscle pyruvate kinase (2.6 mg/ml) was incubated at 25° C with [^{14}C]oADP (0.3 mM) in cacodylate buffer (20 mM, pH 7.5): ($\triangle$) in the presence of 25 mM ATP and 27 mM MgSO$_4$; ($\bigcirc$) in the presence of 1 mM MgSO$_4$. At the indicated times 50 microliter aliquots were taken and placed immediately on a glass fiber disc which was put in cold 10% trichloroacetic acid, washed as described by Corbin & Reimann (1974) and then counted in a scintillation counter. Bezares *et al.* (1987).

muscle pyruvate kinase labeled with trinitrobenzenesulfonate (Johnson *et al.* 1979), and the ε-trinitrophenylated lysine at position 25 is the same lysine found in the oADP-labeled peptide. The same sequence is found between residues 341 and 374 in chicken muscle pyruvate kinase (Lonberg & Gilbert 1983). Regions of high homology to this peptide are also found in yeast pyruvate kinase between residues 313–337 (Burke *et al.* 1983), so as in residues 355–388 of the rat liver enzyme (Lone *et al.* 1986).

Our results show that in cases where the reaction between an oxidized nucleotide and an enzyme give a morpholine-like adduct, it is possible to isolate a labeled peptide if precautions are taken to avoid drastic reaction conditions. The label stays bound to the peptide during reduction, carboxy-methylation, trypsin treatment, and chromatography in NH$_4$HCO$_3$ at pH 8.0, although it is lost during the sequencing process.

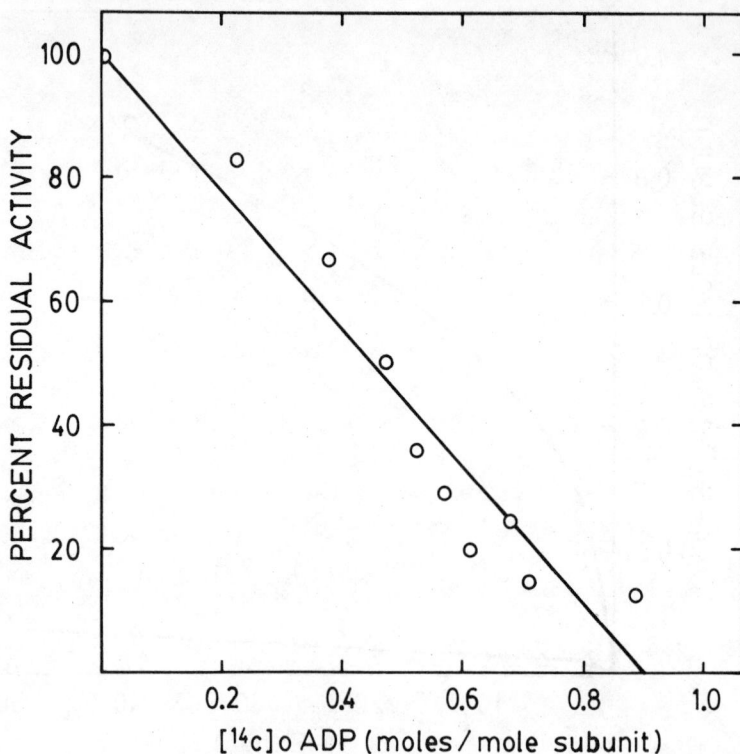

Fig 3.2 — Correlation between inactivation of pyruvate kinase and incorporation of [^{14}C]oADP. Pyruvate kinase (2.6 mg/ml) was incubated at 25° C with [^{14}C]oADP (0.3 mM) in the presence of cacodylate buffer (20 mM, pH 7.5), and MgSO$_4$ (1.0 mM). At intervals, aliquots were withdrawn in order to follow the time-course of the reaction: 10 microliters were used for enzyme activity measurements, and 50 microliters were taken for [^{14}C]oADP incorporation measurement acording to the method of Corbin & Reimann (1974). Bezares *et al*. (1987).

When using oxidized nucleotides it should be kept in mind that these derivatives are not very stable. The half-life of oAMP at 20° C and pH 7.0 has been estimated at 35 hours (Favilla & Bayley 1982); decomposition may produce free phosphate and also generate polymeric material. The stability is enhanced at lower pH or by cooling. Lowe & Beechey (1982b) prepared a decomposition product from oATP (which they called 'compound II') by incubating oATP at pH 9.1 and 35° C for 22 hours. Compound II is formed in a β elimination reaction, liberating the tripolyphosphate ion and retaining the dialdehyde group. This compound is as reactive as oATP. In fact, the incubation of compound II with ATPase produced the inactivation of the enzyme at a faster rate than oATP. However, in contrast to oATP, no rate-saturation effect was seen, suggesting that the inactivation mechanism is different (Lowe & Beechey 1982b). Taking these facts into account, it is recommended that dialdehyde derivatives be used freshly prepared, and their purity be always checked by thin-layer chromatography on polyethyleneimine sheets (Easterbrook-Smith *et al.* 1976) or by electrophoresis on cellulose plates (King & Colman 1983). Otherwise, one may be dealing with different compounds.

In summary, the dialdehyde derivatives have been shown to be very useful site-specific reagents for a variety of ATP and NADH utilizing enzymes, and show the potential for more detailed structure-activity studies.

REFERENCES

Annamalai, A.E. & Colman R.F. (1981) *J. Biol. Chem.* **256** 10276–10283.

Bednar, R.A. & Colman, R.F. (1982) *J. Protein Chem.* **1** 203–224.

Bezares, G., Eyzaguirre, J., Hinrichs, M.V., Heinrikson, R.L., Reardon, I., Kemp, R., Latshaw, S. & Bazaes, S. (1987) *Arch. Biochem. Biophys.* in press.

Brocklehurst, K. (1979) *Biochem. J.* **181** 775–778.

Burke, R.L., Tekamp-Olson, P. & Najarian, R. (1983) *J. Biol. Chem.* **258** 2193–2201.

Colman, R.F., Pal, P.K. & Wyatt, J.L. (1977) *Methods Enzymol.* **46** 240–249.

Colman, R.F. (1983) *Ann. Rev. Biochem.* **52** 67–91.

Corbin, J.D. & Reimann, E.M. (1974) *Methods Enzymol.* **38** 287–290.

Dallocchio, F., Negrini, R., Signorini, M. & Rippa, M. (1976) *Biochim. Biophys. Acta* **429** 629–634.

Easterbrook-Smith, B., Wallace, J.C. & Keech, D.B. (1976) *Eur. J. Biochem.* **62** 125–130.

Favilla, R. & Bayley, P.M. (1982) *Eur. J. Biochem.* **125** 209–214.

Glazer, A.N., DeLange, J.R. & Sigman, D.S. (1975) In *Chemical modification of proteins* (Work, T.S. & Work, E., eds) pp. 167–179, North Holland /Elsevier, Amsterdam.

Gregory, M.R. & Kaiser, E.T. (1979) *Archives Biochem. Biophys.* **196** 199–208.

Hinrichs, M.V. & Eyzaguirre, J. (1982) *Biochim. Biophys. Acta* **704** 177–185.

Jacoby, W.B. & Wilchek, M. E. (1977). Affinity labeling, *Methods Enzymol.* Vol 46, Academic Press, New York.

Johnson, S.C., Bailey, T., Becker, R.R. & Cardenas, J.M. (1979) *Biochem. Biophys. Res. Commun.* **90** 525–530.

King, M.M. & Carlson, G.M. (1981) *Biochemistry* **20** 4382–4387.

King, M.M., Carlson, G.M. & Hayley, B.E. (1982) *J. Biol. Chem.* **257** 14058–14065.

King, M.M. & Colman, R.F. (1983) *Biochemistry* **22** 1656–1665.

Kumar, G., Kalva, V.K. & Brodie, A.F. (1979) *J. Biol. Chem.* **254** 1964–1971.

Lonberg, N. & Gilbert, W. (1983) *Proc. Nat. Acad. Sci. USA* **80** 3661–3665.

Lone, Y.-C., Simon, M.-P., Kahn, A. & Marie, J. (1986) *FEBS Letters* **195** 97–100.

Lowe, P.N., Baum, H. & Beechey, R.B. (1979) *Biochem. Soc. Trans.* **7** 1133–1136.

Lowe, P.N. & Beechey, R.B. (1982a) *Bioorg. Chem.* **11** 55–71.

Lowe, P.N. & Beechey, R.B. (1982b) *Biochemistry* **21** 4073–4082.

Marcus, F. & Haley, B.E. (1979) *J. Biol. Chem.* **254** 259–261.

Mehler, A.H., Kim, J-J.P. & Olsen, A.A. (1981) *Archives Biochem. Biophys.* **212** 475–482.

Meloche, H.P. (1967) *Biochemistry* **6** 2273–2280.

Pal, P.K., Wechter, W.J. & Colman, R. (1975) *J. Biol. Chem.* **250** 8140–8147.

Plapp, B.V. (1982) *Methods Enzymol.* **87** 469–499.

Roy, S. & Colman, R.F. (1979) *Biochemistry* **18** 4683–4690.

Roy, S. & Colman, R.F. (1980) *J. Biol. Chem.* **255** 7517–7520.

Savadambal, K.V., Bednar, R.A. & Colman, R.F. (1981) *J. Biol. Chem.* **256** 11866–11872.

Schaefer, H-J., Scheurich, P., Rathgeber, G. & Dose, K. (1980) *Anal. Biochem.* **104** 106–111.

Scheurich, P., Schaefer, H-J. & Dose, K. (1978) *Eur. J. Biochem.* **88** 253–257.

Severin, E.S., Nesterova, M.V., Gulyaev, N.N. & Shylapmikov, S.V. (1976) *Adv. Enzyme Regul.* **14** 407–444.

Shaw, E. (1970) *The enzymes* 3rd edn **1** 91–146.

Taylor, S.S. & Kerlavage, A.R. (1982) *Fed. Proc.* **41** 660.

Wescott, K.R., Olwin, B.B. & Storm, D.R. (1980) *J. Biol. Chem.* **225** 8767–8771.

Wold, F. (1977) *Methods Enzymol.* **46** 3–14.

4

Photoaffinity labeling and photoaffinity crosslinking of enzymes

Dr Hans-Jochen Schäfer, Institut für Biochemie, Johannes-
Gutenberg Universität, D-6500 Mainz, West Germany

PHOTOAFFINITY LABELING

This chapter is divided into two main parts. The first part, presenting the general principles of photoaffinity labeling, is based mainly on review articles and on an excellent laboratory manual (Knowles 1972, Bayley & Knowles 1977, Bayley 1983). The second part demonstrates the application of photoaffinity labeling and photoaffinity crosslinking with a special enzyme performed in our laboratory.

Advantages of photoaffinity labeling

Chemical modification is often applied to the study of interactions of receptor molecules with their ligands in biological systems.

$$\text{receptor} + \text{ligand} \rightleftharpoons [\text{receptor·ligand}]$$

The receptor binds its ligand usually non-covalently at a specific binding site to form a receptor·ligand complex. Receptors are generally proteins or protein conjugates (e.g. glycoproteins) like enzymes, immunoglobulins, and receptors for hormones, neurotransmitters, or drugs. The corresponding ligands are enzyme substrates, cofactors, allosteric effectors, antigens, hormones, neurotransmitters, or drugs. In contrast to the receptors these ligands differ widely in their molecular structure. Among them are all products of cellular metabolism like sugars, amino acids, nucleotides, and oligomers of these compounds and synthetic products like drugs.

The elucidation of the ligand binding site often allows conclusions about the structure and the function of the receptor protein. One approach to the characterization of a ligand binding site is chemical modification. The first successful results in modifying receptor proteins have been obtained by application of group-specific reagents (Fig. 4.1) (Means & Feeney 1971,

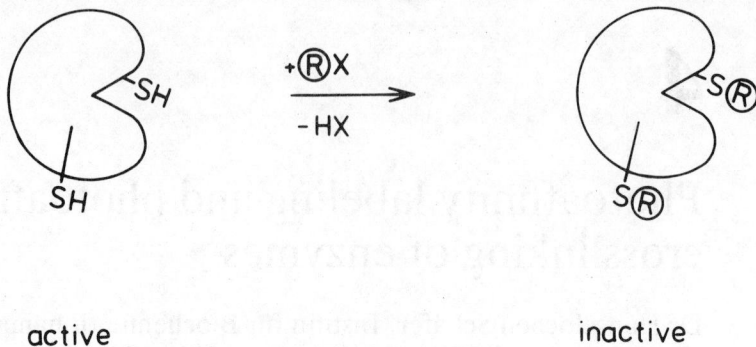

Fig. 4.1 — Modification of proteins by group-specific reagents (®X).

Glazer *et al.* 1975). These substances react specifically with definite amino acid residues of the protein, ideally only with one distinct residue. Group-specific reagents normally attack only nucleophilic groups of the receptor protein. In addition to this, group-specific reagents commonly do not discriminate between amino acid residues inside or outside of the ligand binding site. The inactivation of the receptor's function is only a weak indication for the modification of an essential amino acid residue inside the ligand binding site. The inhibition could also be caused by a conformational change of the protein due to a modification distant from the ligand binding site.

The specific labeling of the binding site is favored by a local increase of the modifying reagent at the ligand binding site. This can be achieved by the incorporation of the reactive group-specific reagent into a ligand creating a reactive ligand analog (affinity label). This analog possesses a certain affinity to the receptor protein to allow specific interaction at the ligand binding site. There it can bind covalently to a nucleophilic amino acid residue due to its reactive group (affinity labeling) (Fig. 4.2) (Jacoby & Wilcheck 1977). Unfortunately, affinity labels, analogous to group-specific reagents, react exclusively with nucleophilic amino acids. However, there are often hydrophobic amino acids involved in the binding of ligands which are not labeled by group-specific reagents or affinity labels with conventional functional groups. Furthermore, affinity labels react immediately after their addition to the system inactivating the receptor owing to the formation of covalent bonds. For this reason it is difficult to study the specific biological interactions of these analogs. Owing to the electrophilic character of their reactive groups most of the added affinity labels are very often hydrolyzed by the solvent water before reaching their target.

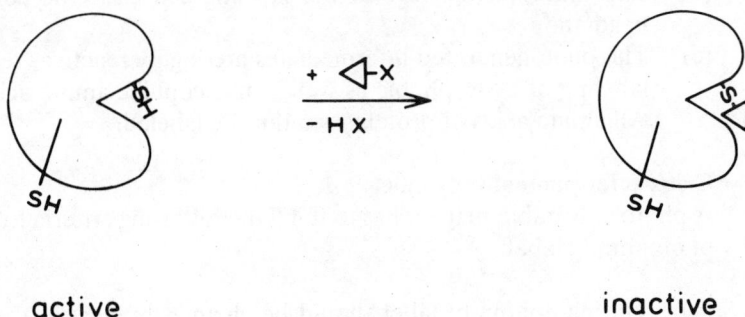

Fig. 4.2 — Affinity labeling of receptor proteins (◁-X: affinity label).

These disadvantages are eliminated when using a non-reactive precursor which can be activated at will. These reagents do not bind covalently to the protein unless activated. Usually photoactivatable ligand analogs are applied for this purpose (photoaffinity labeling) (Fig. 4.3) (Knowles 1972,

Photoaffinity labeling

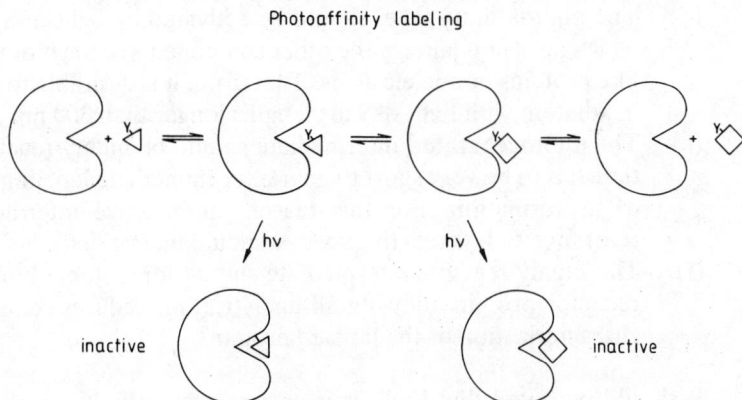

Fig. 4.3 — Photoaffinity labeling of enzymes (◁-Y: substrate analog; ▢-Y: product analog).

Bayley & Knowles 1977, Bayley 1983). Since they are chemically inert in the dark, it is easy to study their biological interactions under these conditions. Upon irradiation of these precursors, highly reactive intermediates are formed which react indiscriminately with all surrounding groups. After its activation, a photoaffinity label interacting at the specific binding site is capable of labeling all amino acid residues of the binding area.

The advantages of photoaffinity labeling are:

(a)   Photoaffinity labels are chemically inert in the dark.

(b)   The time and the rate of the labeling can easily be controlled by irradiation.

(c)   The photogenerated intermediates are highly reactive, resulting in a labeling of hydrophobic as well as nucleophilic amino acid residues. All amino acids of proteins can thus be labeled.

Criteria for photoaffinity labels

A photoactivatable reagent has to fulfill the following criteria to be an ideal photoaffinity label:

(a)   The photoaffinity label should be chemically inert in the dark. All conditions of the labeling experiment, like temperature, pH value and the redox potential of the chemicals present in the mixture, should not damage the photoactivatable precursor.

(b)   It should easily be synthesized with good yields. This criterion is especially important for the synthesis of radiolabeled ligand analogs.

(c)   The chemical changes necessary to make a ligand photoreactive should not alter its conformation too drastically. Modification by bulky photoactivatable compounds may decrease the affinity of the ligand to the receptor protein to an extent where there are no specific biological interactions to allow a specific labeling.

(d)   The photoaffinity label should be activated by light of a wavelength which does not damage the other components of the biological system like proteins or nucleic acids. Therefore, it is desirable to perform the irradiation with light of wavelengths longer than 300 nm.

(e)   The photogenerated intermediate should be highly reactive. Its life-time has to be very short to guarantee immediate labeling at the place of its formation. For this reason, all reactive intermediates that rearrange to less reactive species should be avoided.

(f)   The highly reactive intermediate should form stable bonds with the receptor protein allowing all analytical procedures required for the characterization of the labeled protein.

It should be noted that there is nearly no photoaffinity label which completely fulfills each of these criteria.

Photogenerated reactive intermediates

Two types of reactive intermediates can be generated by irradiation of relatively inert precursors. The first group is produced by the homolytic cleavage of a single bond resulting in two free radicals or one diradical. The second group involves carbenes and nitrenes. They are generated upon homolytic cleavage of a double bond or two adjacent single bonds. Carbenes contain a divalent carbon and nitrenes a monovalent nitrogen (Fig. 4.4). These intermediates are highly reactive because of their lack of electrons. Free carbon radicals possess only seven, nitrenes or carbenes only six electrons in their outer electron shell. All try vehemently to complete an electron octet.

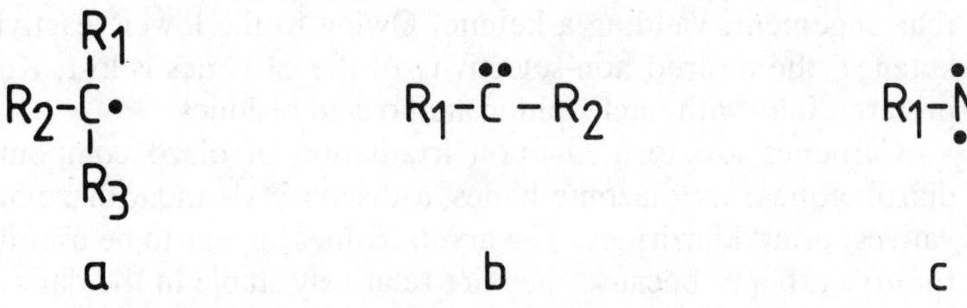

$$R_2-\underset{\underset{R_3}{|}}{\overset{\overset{R_1}{|}}{C}}\bullet$$

$$R_1-\ddot{C}-R_2$$

$$R_1-\ddot{\underset{\cdot\cdot}{N}}$$

a   b   c

Fig. 4.4 — Highly reactive photogenerated intermediates: radical (a), carbene (b), nitrene (c).

Carbenes

A carbene formed by irradiation of diazoacetyl chymotrypsin was the first reactive intermediate applied in photoaffinity labeling (Singh *et al.* 1962). Generally, carbenes can react with polar residues as well as with hydrophobic residues. For this reason carbenes can be applied not only for labeling active sites of enzymes, which usually contain nucleophilic amino acid residues, but also to hydrophobic regions of both receptor binding sites and membrane-integrated proteins. Carbenes react with nucleophilic groups of alcohols or amines yielding, for example, an ether or a secondary amine. Furthermore, they either directly insert into C–H bonds or first abstract a hydrogen atom from the C–H bond forming two radicals which may then combine to form the same product obtained by direct insertion. A fourth important reaction is the addition of carbenes to multiple bonds, yielding a three-membered ring. Addition to an aromatic ring often results in a ring expansion. The reactions of carbenes are analogous to the reactions of nitrenes (cf. Fig. 4.5).

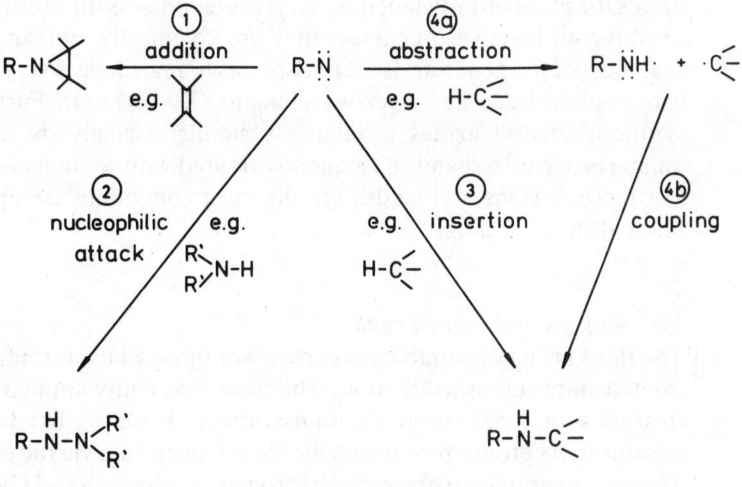

Fig. 4.5 — Reactions of nitrenes.

A disadvantage of carbenes is their susceptibility for rearrangements to less reactive intermediates. Especially, α-diazocarbonyl compounds form α-ketocarbenes upon irradiation which easily undergo intramolecular Wolff rearrangement, yielding a ketene. Owing to the lower reactivity of these ketenes, the desired non-selectivity of the carbenes is lost. Ketenes react preferentially with nucleophilic amino acid residues.

Carbenes are formed upon irradiation of diazo compounds like α-diazoketones, aryldiazomethanes, α-diazoacetyl- and α-diazomalonyl-derivatives, or aryldiazirines. The aryldiazirines appear to be useful precursors for arylcarbenes because they are relatively stable in the dark and are not very susceptible to intramolecular rearrangement after photolysis.

Nitrenes
Nitrenes, the nitrogen analogs of carbenes, were first applied in photoaffinity labeling of antibodies (Fleet *et al.* 1969). Their reactions are similar to those of carbenes (Fig. 4.5). Possible reactions of nitrenes include cycloadditions to multiple bonds forming three-membered cyclic imines (1), addition to nucleophiles (2), direct insertion into C–H bonds yielding secondary amines (3), and hydrogen atom abstraction followed by coupling of the formed radicals to a secondary amine (4a, b). The reactivity of nitrenes, however, is much lower than that of carbenes. They discriminate much more between primary, secondary, and tertiary C–H bonds. In addition, nitrenes are more electrophilic resulting in a preferential attack of O–H bonds over C–H bonds.

Nitrenes can be obtained by photolysis of several azido compounds. Alkyl azides, acyl azides, or aryl azides are possible precursors of nitrenes. Acyl and alkyl azides are not very suitable for photoaffinity labeling because of their instability, their susceptibility for rearrangements, and their improper absorption characteristics. Practical considerations favor aryl azides for photoaffinity labeling. Aryl azides possess three criteria which are ideal for an inactive precursor: they are chemically inert in the dark, the readiness of the generated reactive species to rearrange is very low, and they can be photolyzed at longer wavelengths ($\lambda > 300$ nm). Furthermore, the synthesis of aryl azides is relatively simple. Usually the corresponding aniline is diazotized and subsequently treated with sodium azide at $-20°C$. For these reasons aryl azides are the most commonly used precursors for photoaffinity labeling.

Free radicals and excited states
The third group of suitable precursors are those which form free radicals or excited states upon irradiation. The most frequently applied precursors of this class of reagents are α,β-unsaturated ketones. Irradiation of α,β-unsaturated ketones produces a diradical triplet state via the excited singlet. The radical prefers to abstract a hydrogen atom from a C–H bond, resulting in the formation of two monoradicals which subsequently couple. O–H bonds are much less attacked by the reactive intermediates. For this reason, radicals have advantages over nitrenes or carbenes, for these tend to react preferentially with the solvent water.

Besides the irradiation of α,β-unsaturated ketones, the photoactivation of aryl halides, nitroaryl compounds, purines, pyrimidines, or psoralens can also produce free radicals and excited states.

Photoaffinity labeling, pseudophotoaffinity labeling, and unspecific photolabeling

Irradiation in the presence of a photoaffinity label can lead to the labeling of different regions of the receptor protein.

Ideally, the photoaffinity label binds non-covalently at its binding site. Upon irradiation it forms a covalent bond to an amino acid at the binding site of the receptor protein (specific labeling).

In addition, the photoaffinity label is activated on the way to or from the specific binding site. In this case it labels the protein near the binding site (pseudoaffinity labeling). Pseudoaffinity labeling occurs also if the generated reactive intermediate possesses a longer lifetime. After its generation at the specific binding site, the intermediate dissociates from its place of origin and labels the protein away from this site.

Furthermore, the photoaffinity label may interact non-specifically with the receptor protein distant from its specific binding site. Upon irradiation it will label the protein at this site (unspecific labeling). Unspecific labeling may be important, especially in cases of extremely long lifetimes of the reactive intermediates.

Most of the photoaffinity label reacts after its activation with the solvent water. Normally this product can be easily separated from the labeled protein owing to its lower molecular weight.

The main problems of photoaffinity labeling are pseudophotoaffinity labeling and unspecific labeling. The proportion of pseudophotoaffinity labeling and unspecific labeling should be as low as possible. It is desirable to determine this portion quantitatively. The addition of scavengers reduces nonspecific labeling and pseudophotoaffinity labeling. A scavenger is capable of trapping reactive intermediates outside the ligand site (i.e. those formed outside and those that dissociate from the binding site after their formation). There are two independent means of measuring the degree of unspecific labeling:

(1) A comparison of the results obtained by irradiation in the presence of the photoaffinity label and irradiation in the presence of a non-specific interacting photoactivatable reagent yields the proportions of specific and unspecific labeling. This reagents should possess, as much as possible, structural elements of the photoaffinity label and its photoactivatable group. In contrast to the specific label, it should display no affinity for the ligand binding site of the receptor. It should therefore produce only unspecific labeling.

(2) The second way to discriminate between specific and nonspecific labeling is by using a protecting reagent (Fig. 4.6). For this purpose, either the natural ligand or a non-reactive ligand analog is added in a labeling experiment. The photoaffinity label is displaced from the binding site by

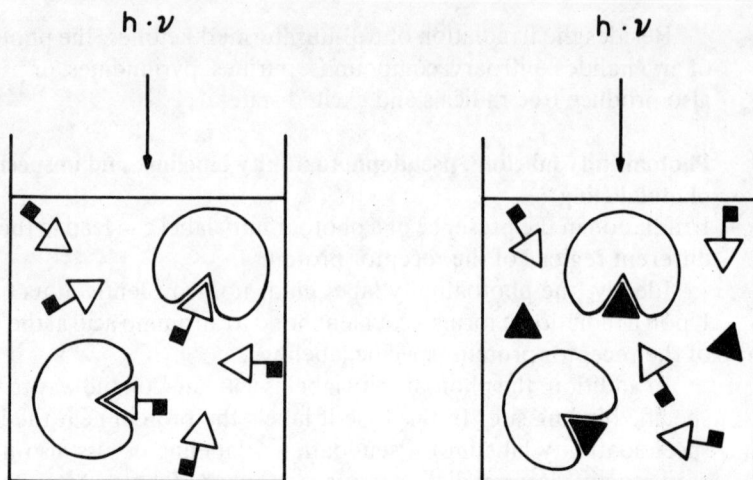

Fig. 4.6 — Unprotected (left) and protected (right) photoaffinity labeling experiments (◁–■: photoaffinity label; ▲: ligand analog).

the ligand or its analog, resulting in the decrease or ideally in the supression, of specific labeling. The comparison between the unprotected and the protected labeling experiments indicates the degree of specific and unspecific labeling.

Examples of photoaffinity labeling
Photoaffinity labeling can be applied to the study of different problems. Three examples are given here.

Identification of a receptor in a mixture
Most hormone receptors are only present in small amounts in the plasma membrane among other proteins. These small amounts can be detected and isolated after photoaffinity labeling with radiolabeled photoactivatable hormone analogs as demonstrated for the insulin receptor (Jacobs *et al.* 1979).

Identification of the ligand binding component of a multisubunit system
In multisubunit protein complexes the subunit containing a specific binding site can be identified by photoaffinity labeling. For example, the nucleotide binding sites of the ATP synthase complex (Vignais & Lunardi 1985) and specific binding sites of ribosomal proteins have been photolabeled and thus characterized (Hsiung *et al.* 1974, Hsiung & Cantor 1974).

Identification and characterization of a ligand binding site within a polypeptide
After photoaffinity labeling and degradation of the receptor protein by proteolysis or chemical cleavage, a small peptide containing the bound photoaffinity label can be isolated and precisely mapped. Ideally, only one

amino acid of the ligand binding site should be labeled. Owing to the high reactivity of the intermediate, however, often several amino acid residues are labeled. 8-Azido ATP, for example, labels three different amino acids of the nucleotide binding site at the β subunit of the mitochondrial F_1ATPase (Hollemans *et al.* 1983).

Photoaffinity crosslinking

The application of bifunctional reagents instead of monofunctional ones results in the crosslinking of proteins. This crosslinking may occur inter- or intramolecularly. The stabilization of the tertiary structure of proteins, the determination of distances between reactive groups in proteins and the study of protein–protein interactions are useful applications of crosslinking reagents. A very important application of bifunctional reagents is for investigating the spatial arrangement of components in biological systems like membranes, ribosomes, or multisubunit enzymes. Crosslinking can also be achieved by reagents with one photosensitive and one conventional functional group, or with two photosensitive groups (photo crosslinking). The introduction of two highly reactive functional groups into a biological ligand creates a tool for studying the vicinity of the specific binding site by crosslinking (affinity crosslinking). A biological ligand possessing two photoactivatable groups forms a photoaffinity label capable of specifically crosslinking proteins upon light activation (photoaffinity crosslinking) (Schäfer 1986). Photoaffinity crosslinking possesses the same advantages over crosslinking or affinity crosslinking as photoaffinity labeling over protein modification by group-specific reagents or affinity labels.

The photoaffinity labeling experiment

The following experimental sequence can be recommended.

Demonstration of the biological activity of the photoaffinity label

The most important precondition for a useful photoaffinity label is its biological activity. The photoaffinity label must show the same specific interactions with the receptor protein as the natural ligand. A suitable photoaffinity label for an enzyme should be a substrate (in the dark) or at least a competitive inhibitor.

Light-induced inactivation in the presence of the photoaffinity label

Irradiation of the receptor protein in the presence of the photoaffinity label should decrease or even destroy the biological activity of the receptor protein. This inactivation should not be observed in the following control experiments: neither by incubation of the receptor with the photoaffinity label in the dark (dark control) nor by irradiation of the receptor protein in the absence of the photoaffinity label (light control).

Protection from light-induced inactivation by addition of natural ligands
The addition of the natural ligand or a non-reactive ligand analog should protect the receptor protein against the attack of the photoaffinity label, resulting in the maintenance of biological activity. Substrates, products, or competitive inhibitors are suitable protecting reagents for an enzyme.

Photoaffinity labeling with radioactive photoactivatable analogs
If the preceding experiments have been successful in demonstrating the specific interaction of the receptor protein with the photoaffinity label and the specific photoinactivation of the receptor protein, it is useful to employ a radiolabeled photoaffinity label. After the incorporation of the photoaffinity label, the degree of labeling can be measured, the labeled subunits of protein complexes can be identified, and essential amino acid residues or the sequence of the ligand binding site can be determined.

Protection from photoaffinity labeling
The addition of the natural ligand or a ligand analog should decrease or even prevent the labeling of the receptor protein in analogy to the protection from light-induced inactivation.

PHOTOAFFINITY LABELING AND PHOTOAFFINITY CROSSLINKING OF F₁ATPase from *Micrococcus luteus*

The second part of this chapter demonstrates a special example of photo-affinity labeling performed in our laboratory. We have synthesized 8-azido ATP (8-N_3ATP), the fluorescent 8-azido-1,N^6-etheno ATP (8-$N_3\varepsilon$ATP), and the bifunctional (crosslinking) 3'-arylazido-β-alanine-8-azido ATP (DiN_3ATP) to characterize the nucleotide binding site of bacterial F₁ATPases (Fig. 4.7) (Schäfer *et al.* 1978a, 1978b, 1980a). F₁ATPase is the catalytic portion of the ATP synthase complex (Senior & Wise 1983). This complex is the terminal enzyme in oxidative phosphorylation and photo-phosphorylation. It uses the electrochemical potential energy of a proton gradient to synthesize ATP from ADP and phosphate. The ATP synthase complex is composed of a water-soluble component (F₁ATPase) and a membrane-integrated part (F₀) (Fig. 4.8). Solubilized F₁ATPase, generally consisting of five different subunits (α, β, γ, δ and ε), contains the catalytic sites for ATP synthesis and ATP hydrolysis. The stoichiometry of F₁ATPases is $\alpha_3\beta_3\gamma\delta\varepsilon$. Photoaffinity labeling and photoaffinity crosslinking are demonstrated with a bacterial F₁ATPase from *Micrococcus luteus* (Scheurich *et al.* 1978, Schäfer *et al.* 1980b, Schäfer & Dose 1984).

Biological activity of 8-azido-adenosine nucleotides
The 8-substituted ATP analogs (8-BrATP, 8-N_3ATP, 8-$N_3\varepsilon$ATP, DiN_3ATP) are hydrolyzed by F₁ATPase from *Micrococcus luteus* in the presence of divalent metal ions. The rate of hydrolysis for these analogs is drastically decreased compared with the natural substrate ATP. This is due to the changed conformation of 8-substituted ATP analogs which prefer the

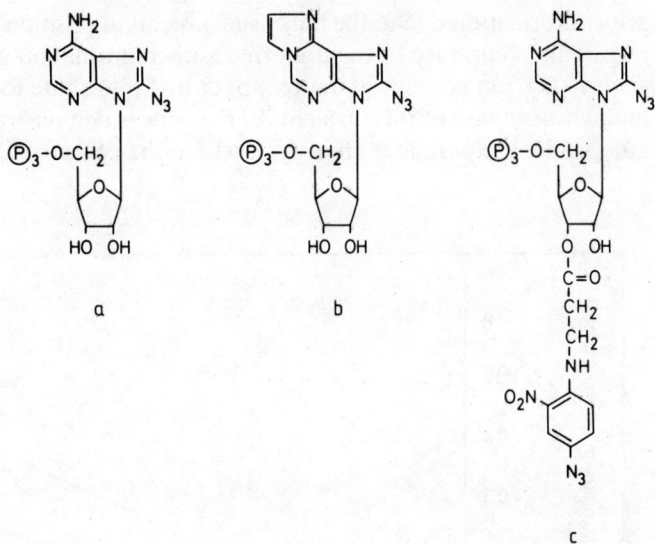

Fig. 4.7 — Photoactivatable ATP analogs: $8\text{-}N_3ATP$ (a); $8\text{-}N_3\varepsilon ATP$ (b); diN_3ATP (c).

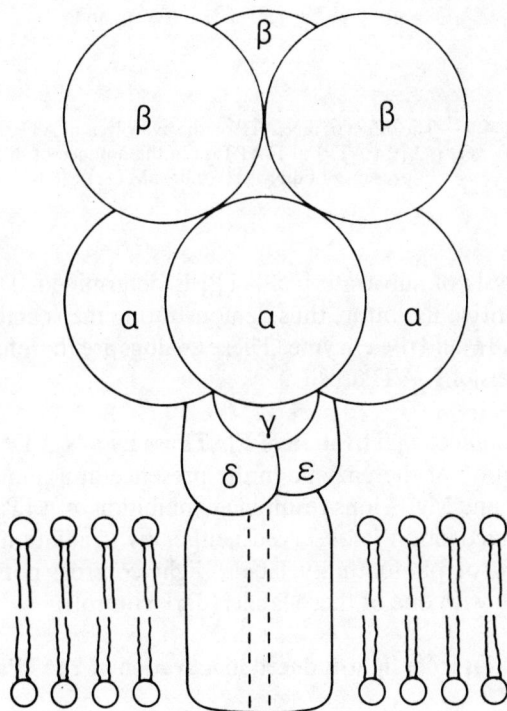

Fig. 4.8 — Possible structure of the ATP synthase complex.

syn-conformation due to the bulky substituents at position 8 of the adenine ring. This is contrary to the preferred anti-conformation of ATP (Ikehara *et al.* 1972). On account of the very poor hydrolysis rate for DiN$_3$ATP, the biological activity of this divalent ATP analog is demonstrated by a second independent experiment (Fig. 4.9), where the effect of DiN$_3$ATP on the

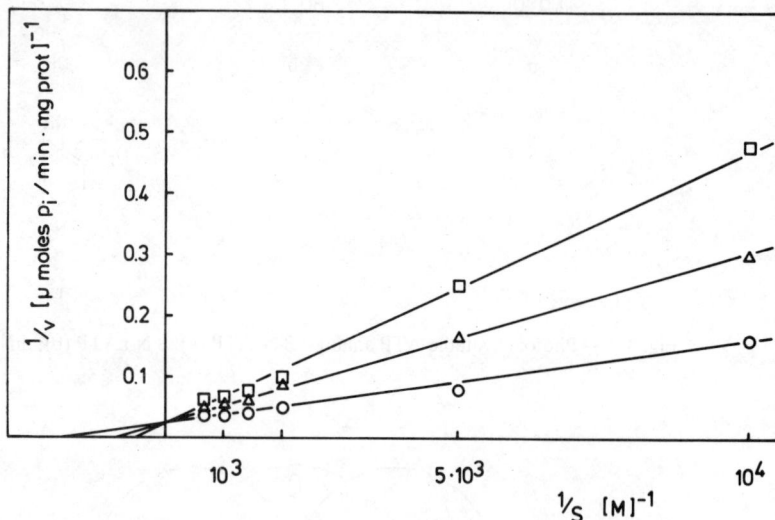

Fig. 4.9 — The effect of diN$_3$ATP on the hydrolysis of ATP. Lineweaver–Burk plots : 1/v versus 1/[Ca·ATP] of F$_1$APTase in the absence of di N$_3$ATP (○) and in the presence of diN$_3$ATP (0.05 mM (△)), (0.075 mM (□)).

hydrolysis of substrate [Ca·ATP] is determined. DiN$_3$ATP behaves as a competitive inhibitor, thus demonstrating the specific interaction between DiN$_3$ATP and the enzyme. These analogs are therefore suitable photoaffinity labels of F$_1$ATPases.

Light-induced inactivation of F$_1$ATPase by 8-N$_3$ATP analogs
Irradiation of the enzyme in the presence of a photoactivatable 8-N$_3$ATP analog and Mg^{2+} ions result in an inhibition of ATPase activity (Fig. 4.10). This inactivation is observed neither by irradiation of the enzyme in the absence of photoaffinity labels (light control) nor by incubation of the enzyme with one of these labels (dark control).

Protection from light-induced inactivation of F$_1$ATPase by addition of ATP or ADP
The addition of ATP or ADP to the photoaffinity labeling experiment prior to irradiation protects the F$_1$ATPase from light-induced inhibition (Fig. 4.11). These nucleotides compete with the photoaffinity labels for nucleo-

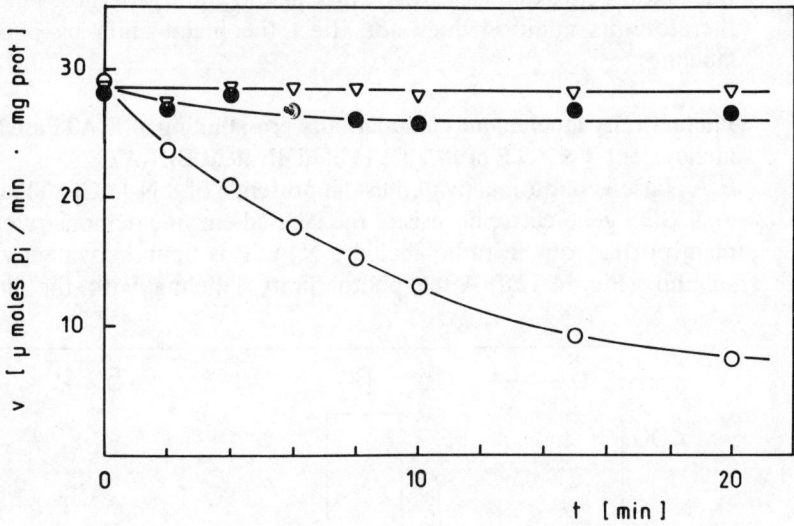

Fig. 4.10 — Light-induced inhibition of F_1ATPase. Irradiation in the presence of 0.5 mM Mg·diN₃ATP (○); light control in the absence of diN₃ATP (▽); dark control in the presence of 0.5 mM Mg·diN₃ATP (●).

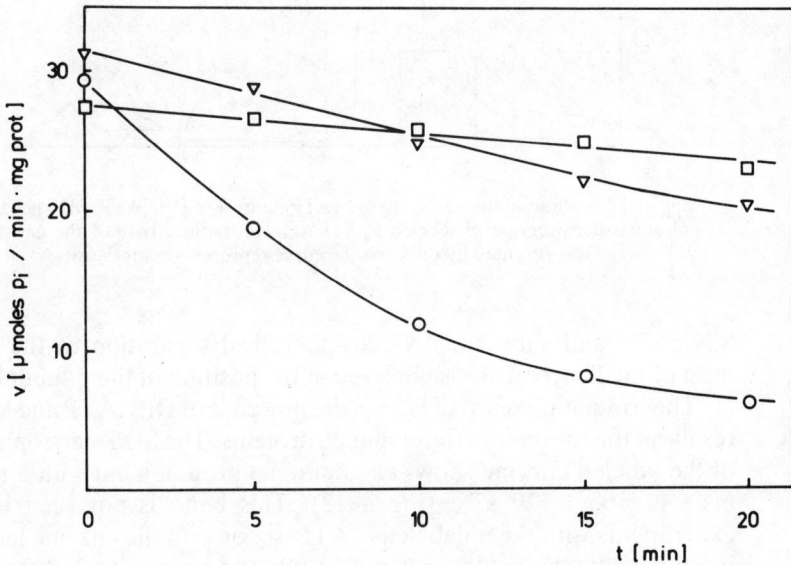

Fig. 4.11 — The effect of added Mg·nucleotides (1 mM Mg·AMP (○); 1 mM Mg·ADP (▽); 1 mM Mg·ATP (□)) on the light -induced inhibition of F_1ATPase by 0.5 mM Mg·diN₃ATP.

tide binding sites of the enzyme. AMP has no affinity for these binding sites, therefore its addition does not affect the inactivation by photoaffinity labeling.

Photoaffinity labeling and photoaffinity crosslinking of F₁ATPase by monovalent 8-N₃ATP or 8-N₃εATP and divalent DiN₃ATP

F_1ATPase is photoinactivated in the presence of 8-N$_3$[^{14}C]ATP and Mg^{2+} ions. SDS gel electrophoresis of the labeled enzyme demonstrates that the main portion of the radiolabeled 8-N$_3$ATP is bound covalently to the β subunit (Fig. 4.12). After photoaffinity labeling with the fluorescent

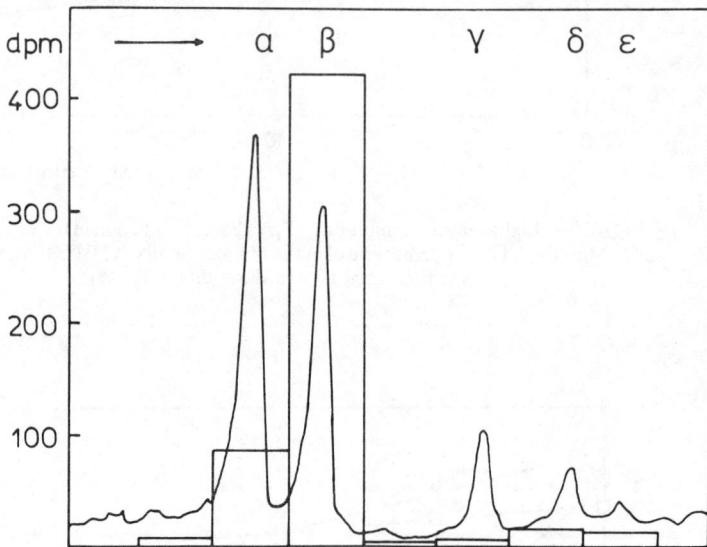

Fig. 4.12 — Photoaffinity labeling of F_1ATPase by 8-N$_3$[^{14}C]ATP. SDS gel electrophoresis densitogram of labeled F_1ATPase. The radioactivity of the gel slices is represented by the bars; the curve represents color density.

8-N$_3$εATP and subsequent electrophoretical separation of the subunits, most of the fluorescence is observed at the position of the β subunit as well.

The irradiation of F_1ATPase in the presence of DiN$_3$ATP and Mg^{2+} ions results in the formation of crosslinked proteins. The SDS electrophoresis gel of the labeled enzyme shows an additional protein band with a molecular mass of about 120 kDa (Fig. 4.13). This band is not seen in control experiments with the unlabeled F_1ATPase, or with the enzyme labeled and inactivated by the monovalent 8-N$_3$ATP. The composition of the crosslink was elucidated by hydrolytic cleavage of the crosslink and subsequent SDS gel electrophoresis (Fig. 4.14). The crosslink is almost entirely split into two protein bands of nearly identical concentration with a mobility corresponding to that of the α and β subunits. These data indicate a subunit composition α-β for the crosslink (Fig. 4.15).

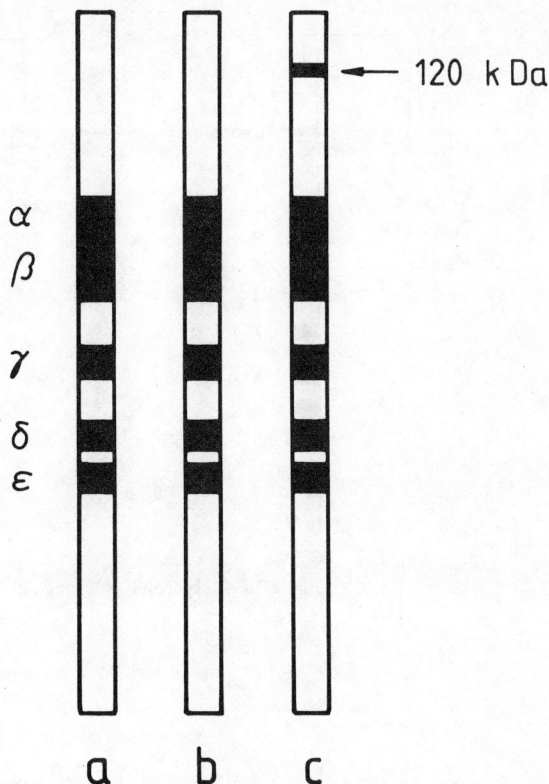

Fig. 4.13 — Photoaffinity crosslinking of F_1ATPase. SDS electrophoresis gels of labeled (crosslinked) F_1ATPase: native F_1ATPase (control) (a); F_1ATPase labeled by monovalent 0.5 mM Mg·8-N_3ATP (b); F_1ATPase labeled by divalent 0.5 mM Mg·diN$_3$ATP (c).

Protection from photoaffinity labeling and photoaffinity crosslinking of F_1ATPase by addition of ATP or ADP

The addition of ATP or ADP prior to labeling protects the F_1ATPase from photoaffinity labeling with 8-N_3[^{14}C]ATP (Table 4.1) or from photoaffinity crosslinking by DiN$_3$ATP. Both effects are not influenced by addition of AMP. These findings demonstrate that the hydrolytic nucleotide binding sites are specifically labeled and involved in the specific formation of the crosslinks. A further evidence for the specific labeling of the F_1ATPase by 8-N_3ATP is obtained by a control experiment with 8-N_3AMP. 8-N_3AMP does not interact specifically with hydrolytic nucleotide binding sites of the enzyme. For this reason, 8-N_3AMP labels the F_1ATPase only unspecifically and therefore the amount of labeling and light-induced inactivation is drastically reduced.

As a consequence of these experiments, it can be concluded that the catalytic nucleotide binding sites of F_1ATPase from *Micrococcus luteus* are

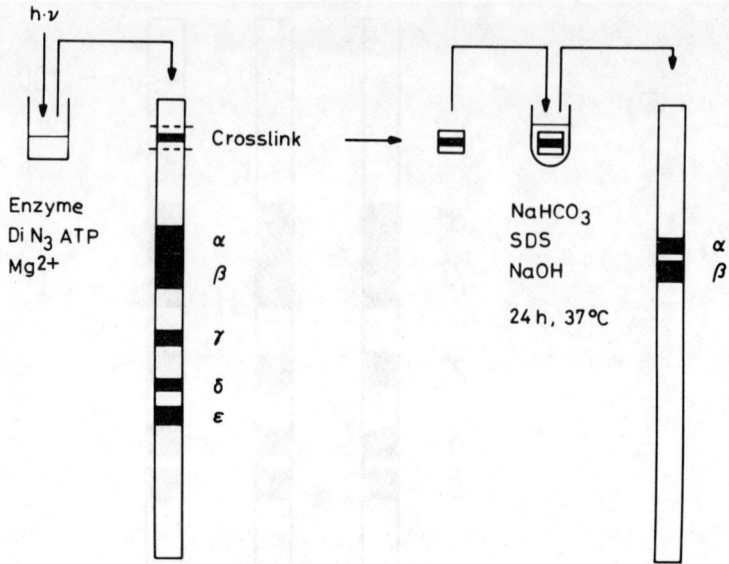

Fig. 4.14 — Hydrolytic cleavage of the crosslink.

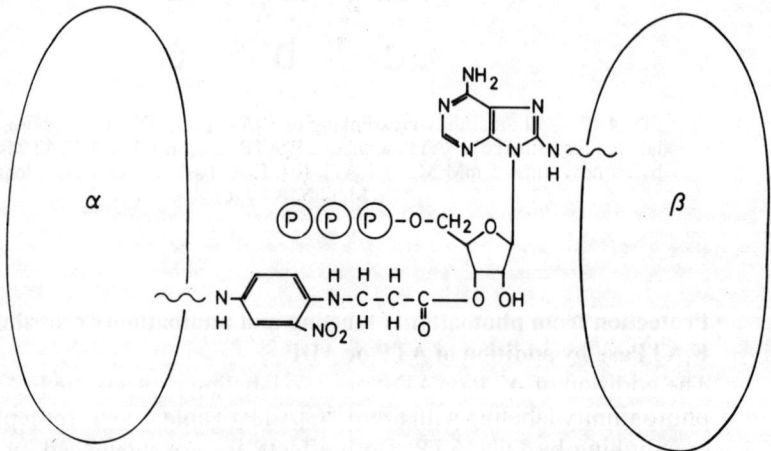

Fig. 4.15 — Possible structure of the α-β crosslink.

located on the β subunits very closely to the α subunits, probably at the interfaces between them.

Further characterization of the binding site can be performed, for example, by mapping the site and determining the nature of the labeled amino acid residues, as has been shown for the mitochondrial F_1ATPase by Hollemans *et al.* (1983).

Table 4.1
Influence of various effectors on photoaffinity labeling of the α and β subunits with 8-N$_3$[^{14}C]ATP and 8-N$_3$[^{14}C]AMP.

Label	Effector	dpm α	β	Inactivation (%)
8-N$_3$ATP	Mg^{2+}	83	386	80
8-N$_3$ATP	Mg^{2+}, AMP	68	297	78
8-N$_3$ATP	Mg^{2+}, ADP	59	53	17
8-N$_3$ATP	Mg^{2+}, ATP	27	32	21
8-N$_3$AMP	Mg^{2+}	78	35	12

ACKNOWLEDGEMENTS

The author appreciates the help of G. Rathgeber, M. Mittelmann-Sicurella and Dr K. Dose, Universität Mainz. The studies were supported by grants from the Deutsche Forschungsgemeinschaft and the Alexander von Humboldt-Stiftung.

REFERENCES

Bayley, H. & Knowles, J.R. (1977) *Methods Enzymol.* **46** 69–114.
Bayley, H. (1983) *Laboratory Techniques in Biochemistry and Molecular Biology*, Vol. 12: Photogenerated Reagents in Biochemistry and Molecular Biology (Work, T.S. & Burdon, R.H., eds), Elsevier, Amsterdam.
Fleet, G.W.J., Porter, R.R. & Knowles, J.R. (1969) *Nature* **224** 511–512.
Glazer, A.N., DeLange, R.J. & Sigman, D.S. (1975) *Laboratory Techniques in Biochemistry and Molecular Biology*, Vol. 4, Part 1: Chemical Modification of Proteins (Work, T.S. & Work, E., eds), North-Holland, Amsterdam.
Hollemans, M., Runswick, M.J., Fearnley, I.M. & Walker, J.E. (1983) *J. Biol. Chem.* **258** 9307–9313.
Hsiung, N. & Cantor, C.R. (1974) *Nucleic Acids Res.* **1** 1753–1762.
Hsiung, N., Reines, S.A. & Cantor, C.R. (1974) *J. Mol. Biol.* **88** 841–855.
Ikehara, M., Uesugi, S. & Yoshida, K. (1972) *Biochemistry* **11** 830–836.
Jacobs, S., Hazum, E., Shechter, Y. & Cuatrecasas, P. (1979) *Proc. Natl. Acad. Sci. USA* **76** 4918–4921.
Jacoby, W.B. & Wilcheck, M. (1977) *Methods Enzymol.* **46**
Knowles, J.R. (1972) *Acc. Chem. Res.* **5** 155–160.
Means, G.E. & Feeney, R.E. (1971) *Chemical Modification of Proteins*, Holden-Day Inc., San Francisco.
Schäfer, H.-J. (1986) *Methods Enzymol.* **126** in press.

Schäfer, H.-J., Scheurich, P. & Dose, K. (1978a) *Liebigs Ann. Chem.* **1978** 1749–1753.

Schäfer, H.-J., Scheurich, P., Rathgeber, G. & Dose, K. (1978b) *Nucleic Acids Res.* **5** 1345–1351.

Schäfer, H.-J., Scheurich, P., Rathgeber, G., Dose, K., Mayer, A. & Klingenberg, M. (1980a) *Biochem. Biophys. Res. Commun.* **95** 562–568.

Schäfer, H.-J., Scheurich, P., Rathgeber, G. & Dose, K. (1980b) *Anal. Biochem.* **104** 106–111.

Schäfer, H.-J. & Dose, K. (1984) *J. Biol. Chem.* **259** 15301–15306.

Scheurich, P., Schäfer, H.-J. & Dose, K. (1978) *Eur. J. Biochem.* **88** 253–257.

Senior, A.E. & Wise, J.G. (1983) *J. Membr. Biol.* **73** 105–124.

Singh, A., Thornton, E.R. & Westheimer, P.C. (1962) *J. Biol. Chem.* **237** 3006–3008.

Vignais, P.V. & Lunardi, J. (1985) *Ann. Rev. Biochem.* **54** 977–1014.

5

Sensitized photo-oxidation of amino acids in proteins

Dr Eduardo Silva, Laboratorio de Química de Proteínas y Alimentos, Facultad de Química, Universidad Católica de Chile, Santiago, Chile

The role of enzymes as catalysts of the reactions occurring in living organisms as well as their increasing use in biotechnology have stimulated the investigations devoted to provide information regarding the molecular details of the active site, as an approach to the understanding of their mode of action. The chemical modification method has been traditionally used to establish the amino acid residues that participate in the active site of enzymes, and the sensitized photo-oxidation of amino acids in proteins can be included as a particular case of this method.

The photodynamic action or sensitized photo-oxidation represents a process which originates a chemical change in the substrate by irradiation of the system under study with visible light in the presence of a sensitizer. The photodynamic action requires molecular oxygen, and the sensitizer in an excited triplet state seems to be the species responsible for the beginning of the reaction.

MECHANISM OF THE SENSITIZED PHOTO-OXIDATION

Two mechanisms have been proposed to explain sensitized photo-oxidation (Foote 1968, Kramer & Maute 1972, Nilsson & Kearns 1973). In the type I mechanism, the substrate initially reacts with the sensitizer in the triplet state and then with molecular oxygen. In the type II mechanism, the excitation energy is transferred from the sensitizer in the triplet state to the molecular oxygen, giving rise to $^1O_2$, which reacts with the substrate. In some cases, the two mechanisms can be present, depending on the sensitizer used (Kramer & Maute 1972, Silva 1979, Sconfienza et al. 1980, Tsai et al. 1985). It is also possible to initiate a photo-oxidative process by means of the enzymatic excitation of the sensitizer through a chemical reaction, in a process called 'photochemistry without light' (Durán et al. 1983).

Of the two types of singlet oxygen, $^1\Delta$ and $^1\Sigma$ (Kasha & Branham 1979), only the first species has an effective lifetime in aqueous systems (Rodgers & Snowden 1982, Ogilby & Foote 1983). One of the methods used to demonstrate the participation of the $^1O_2$ in biological and chemical processes consists in the electronic relaxation of their excited states by the action of chemical agents. Hasty *et al.* (1972) reported that sodium azide produces a significant decrease in the lifetime of the singlet oxygen; this was used by Nilsson *et al.* (1972) to demonstrate the participation of $^1O_2$ in the sensitized photo-oxidation of the amino acids and in the photodynamic inactivation of alcohol dehydrogenase.

The measurement of chemical luminescence at a specific wavelength is frequently used as reliable evidence for the presence of $^1O_2$ in biological reactions (Cadenas *et al.* 1980). The emission of singlet oxygen dimol (chemoluminescence arising from simultaneous transitions in pairs of singlet oxygen molecules) takes place at 634 and 703 nm (Khan & Kasha 1970). For the monomolecular species, the emission is observed at 1,268 and 1,407 nm (Khan & Kasha 1979).

Tertiary aliphatic amines have also been described as quenchers of the $^1O_2$ (Ouannes & Wilson 1968). Another useful diagnostic method to test for the presence of singlet oxygen is based on the effect of deuterated water over the lifetime of the $^1O_2$ (Merkel *et al.* 1972). The lifetime of $^1O_2$ has a value of 53–68 μs in D_2O (Linding & Rodgers 1979, Ogilby & Foote 1983), and of 4 μs in H_2O (Rodgers & Snowden 1982). Thus the reaction rate, neglecting solvent isotope effects for photo-oxidation, is increased by a factor of 15 in going form H_2O to D_2O.

Just as the presence of $^1O_2$ can be considered as evidence for a type II mechanism, the type I mechanism is suggested by the quenching effect of the electron acceptor $[Fe(CN)_6]^{-3}$ (Dewey & Stein 1970, Rossi *et al.* 1981). Koizumi & Usui (1972) have demonstrated that in an aqueous medium the interaction between two molecules of sensitizer, one in the triplet state and the other in the basal state, leads to the formation of two radical ions (sens.$^+$ and sens.$^-$). These radical ions are then efficiently attacked by molecular oxygen, giving rise to superoxide ions ($O\cdot_2^-$), a species that is highly reactive. Other studies of Zwicker & Grossweiner (1963) have pointed out that the sensitizer in the triplet state reacts with the substrate, producing radical ions, which in turn react with molecular oxygen, leading to the formation of oxidized derivatives of the substrate.

Tsai *et al.* (1985) have demonstrated that for a series of enzymes, the photodynamic process is preceded by the binding between the enzyme and the sensitizers.

SELECTIVITY OF THE PHOTODYNAMIC EFFECT IN PROTEINS

The photochemical modification produced in proteins by the effect of the sensitized photo-oxidation is circumscribed to the specific modification of the side-chains of certain amino acid residues, no alterations being produced at the level of the peptide bonds (Weil *et al*. 1953, Ghiron & Spikes 1965). Out of the twenty types of amino acid residues that can be part of a protein, only the side-chains of Cys, His, Met, Trp and Tyr residues are susceptible of being modified by the effect of sensitized photo-oxidation (Galiazzo *et al*. 1968, Silva *et al*. 1974).

In the case of Cys, photo-oxidation has been studied over a wide range of pH, and the only characterized photoproduct has been cysteic acid (Jori *et al*. 1969).

Tomita *et al*. (1969) reported that the His sensitized photo-oxidation leads to the formation of aspartate and urea, which was subsequently confirmed by Tsai *et al*. (1985).

By means of photo-oxidation, Met is converted into methionine sulfoxide, and after a long reaction period, into methionine sulfone, and apparently into homocysteic acid (Weil *et al*. 1953, Benassi *et al*. 1967, Jori & Cauzzo 1970). When the determination of the amino acid composition of a protein is used with the purpose of quantifying the photo-oxidation of the Met residues present in it, an alkaline hydrolysis must be previously carried out, since methionine sulfoxide is reverted to methionine during an acid hydrolysis (Jori *et al*. 1968a,b, Risi *et al*. 1973).

The products of Trp photo-oxidation have been the subject of numerous studies (Benassi *et al*. 1967, Asquith & Rivett 1971, Creed 1984a). Saito *et al*. (1979) have proposed three pathways starting with a common intermediate (hydroperoxiindolalanine) for the transformation of Trp in its photoproducts kinurenine and N-formylkinurenine, whose fluorescent characteristics have been reported by Fukunaga *et al*. (1982).

Recent work (Nakagawa *et al*. 1985) has revealed that the pH not only exerts an influence on Trp oxidation velocity, but also influences the formation of other photoproducts, such as 5-hydroxy formylkynurenine at pH values > 7.0 and a tricyclic hydroperoxide in the pH range 3.6–7.1.

The Tyr photo-oxidation products have not been characterized yet (Creed 1984b), and only the photoproducts of phenolic derivatives analogous to Tyr have been identified (Saito *et al*. 1975, Matsuura 1977).

The photo-oxidation rate of the susceptible amino acids is strongly dependent on reaction conditions, as has been extensively discussed in the classic work of Spikes & Livingston (1969). At pH values smaller than six, the His and Tyr residues are not photo-oxidizable, owing to protonation of the imidazole ring and because of the weak activation effect of the hydroxylic substituent of the benzene ring (Sluyterman 1962). Trp presents a wide range of photoreactivity; only the photoproducts obtained vary with the pH employed, as was mentioned before.

The methionyl residues can be selectively converted to the sulfoxide by irradiation in acetic acid solution at low temperatures, in the presence of rose bengal or methylene blue as sensitizers (Jori *et al*. 1968a,b); in the presence of hematoporphyrin, methionine can also be selectively photo-oxidized in aqueous solutions buffered at pH values lower than 6.5 (Jori *et al*. 1969).

Crystal violet acts selectively on cysteine both in aqueous solution over the pH range 2.5–9.0 and in 5 to 95% acetic acid (Bellin & Yankus 1968, Jori *et al*. 1969).

The efficiency and selectivity of photosensitized reactions may depend on various factors (Foote 1976, Spikes 1977, Jori & Spikes 1981). Binding (complexation) of sensitizer to the substrate may be one of the most important questions to be answered (Grossweiner 1969, Grossweiner & Kepka 1972, Silva & Gaule 1977, Jori & Spikes 1981, Tsai *et al*. 1985). Most sensitizers can complex reversibly with nucleic acids, proteins, and polysaccharides (Amagasa 1981). Sensitizer-substrate complexation may increase the population of the triplet state and decrease the collisional quenching of the triplet state (Bellin 1968). Furthermore, the vicinity of the substrate and sensitizer may enhance the probability of a direct reaction of these counterparts. Type I mechanism is thus favored. Type I and Type II reactions for important biomolecules have been reviewed (Foote 1976, Spikes 1977).

Tsai *et al*. (1985) have shown that the rates of photo-inactivation are at least an order of magnitude faster with efficient sensitizers (methylene blue, hematoporphyrin, rose bengal, erythrosin B, and eosin Y) than with inefficient ones (fluorescein, acridine orange, riboflavin, and pyridoxal 5'-phosphate). No obvious correlation exists between sensitizing efficiency and dye structures. Although the physicochemical parameters that characterize sensitizing efficiency are yet to be defined, halogenated xanthines and thiazine with low triplet state energies are shown to be effective photosensitizers.

Higher quantum yields for photoinactivation of lysozyme sensitized by riboflavin in relation to those obtained in the presence of methylene blue (Shugar 1952, Hopkins & Spikes 1969, 1970, Silva *et al*. 1974, Silva & Gaule 1977) have been interpreted through the photo-induced complex obtained on irradiating lysozyme in the presence of riboflavin (Silva & Gaule 1977, Ferrer & Silva 1981). In subsequent studies, it was demonstrated that the bond between riboflavin and the enzyme is specific, and that it is produced at the Trp residues present in lysozyme (Ferrer & Silva 1985).

DETERMINATION OF BURIED AND EXPOSED GROUPS IN PROTEINS BY MEANS OF SENSITIZED PHOTO-OXIDATION

In the three-dimensional structure of a protein, the photo-oxidizable amino acid residues can be located either at the surface of the molecule in contact with the solvent or in the interior at the hydrophobic regions. To these two groups, one should add those residues that are in an intermediate situation and therefore are neither completely exposed to the solvent nor correspond to the buried ones in the hydrophobic regions.

It has been found that photo-oxidizable amino acid residues are modified in a different manner depending on their location in the three-dimensional structure of the protein. For example, Weil *et al*. (1965) found that only the His residues were photo-oxidized when they studied insulin in its native form, but the modification was extended to the Tyr residues when 8M urea was added. RNAase suffers sequential photo-oxidation of its Met residues when it is irradiated in the presence of increasing concentrations of acetic acid, using hematoporphyrin as sensitizer (Jori *et al.* 1970). The homologous proteins α-lactalbumin and lysozyme present different reactivity to sensitized photo-oxidation in their native and denatured forms (Edwards & Silva, 1985).

Ray & Koshland (1961, 1962) proposed a kinetic model to calculate the number of residues of amino acids exposed on the surface, and therefore of those located in the hydrophobic interior of the protein. They postulated that the residues of amino acids exposed, owing to a higher accessibility to the reactive species, would be photo-oxidized in a more efficient form than the internal ones. When studies tending to clarify the mechanism of sensitized photo-oxidation were carried out, it was found that in a very large number of cases sensitized photo-oxidation took place via singlet oxygen through a Type II mechanism (Jori & Spikes 1981). This fact not only identified the reactive species participating in sensitized photo-oxidation, but also had implications in the interpretation of the kinetic analysis of Ray & Koshland (1961). In order to be able to apply this model, it was necessary to find an explanation for the difference in reactivity of the internal and external amino acid residues to singlet oxygen. An adequate explanation was not easy with the information available. The lifetime of the singlet oxygen is very much higher in an apolar medium than in a polar solvent like H_2O (Turro 1978), a fact which favors the existence of a reactive species in the hydrophobic interior of the protein. Besides, the diffusion velocity of molecular oxygen is much higher than that of any organic molecule, which would also favor the arrival of singlet oxygen to the hydrophobic interior of the proteins (Turro 1978).

In 1973, Lakowicz & Weber performed a series of experiments in which they studied the intrinsic fluorescence quenching of proteins by the action of molecular oxygen under high pressure. They found that molecular oxygen is able to quench the fluorescence of the Trp residues, including those that are in the interior of the protein structure, which means that oxygen is able to diffuse and reach this type of residue formerly considered to be inaccessible.

In recent years, several investigations provided the necessary information required to understand the behavior of oxygen in the interior of a protein. In 1982, Eftink & Jameson studied fluorescence quenching of alcohol dehydrogenase. This enzyme is characterized by having two Trp residues in its structure: Trp-15 that is exposed to the solvent, and Trp-314 that is in the hydrophobic interior of the protein. These authors found that quenching by oxygen was five times more effective for the Trp-15 residue. According to the most recent evidence available, the value of the quenching constant by oxygen for Trp-314 is the lowest that has been found for proteins. These results support the kinetic analysis of Ray & Koshland (1961) that is based on the fact that the residues of internal amino acids are less accessible to the reactive species (singlet oxygen) than the exposed residues that would be completely accessible and therefore more reactive. This is also in accordance with the work of Calhoun *et al.* (1983a,b), who, using oxygen and other small molecules as fluorescence quenchers, found that all of them were more efficient for the Trp residues accessible to the solvent.

In addition, Reddi *et al.* (1984) found that in spite of the fact that the singlet oxygen has a longer lifetime in solvents less polar than water, the efficiency of photo-oxidation was smaller in apolar media. These results are also supported by the fact that the internal residues are less reactive than the external ones, since they are in a less polar environment.

In summary, the methodology proposed by Ray & Koshland (1961) still retains its validity with the simplifications and limitations characteristic of any kinetic model.

Three modes of kinetic behavior can be expected in sensitized photo-oxidation:

(a)   No effect is observed, in spite of the use of adequate conditions to modify certain amino acid residues. This kind of result indicates that these residues are not accessible to the reactive species or that their reactivity is very low, and therefore they are located in the hydrophobic region of the protein.

(b)   A straight line is obtained in a semilogarithmic plot of the remaining concentration of the amino acid as a function of irradiation time. This indicates that all the residues could be photo-oxidized with the same first-order rate constant (k).

$$\frac{C}{C_O} = e^{-kt} \tag{1}$$

This situation is clearly present when a protein has only one amino acid residue of a certain type and the photo-oxidation phenomenon takes place in this residue. Such is the case of the His-15 residue in lysozyme (Risi *et al.* 1973).

With the purpose of estimating the exposure of the photo-oxidizable residue to the solvent, the rate constant measured experimentally is compared with the one obtained for the amino acid in its free form (Jori *et al.* 1970). It is also possible to resort to small peptides that contain the amino acid under study, or to the same protein in its denatured form (Edwards & Silva 1985).

(c)   A non-linear semilogarithmic plot is obtained, as the one shown in Fig. 5.1 for the photo oxidation of the Trp residues in α-lactalbumin (Edwards & Silva 1985). In this case, the shape of the observed curve for residue modification suggests that some of the Trp react with higher velocity than others. This difference in reactivity is attributable to their location in the structure, and therefore to the polarity of their environment and their accessibility to singlet oxygen, as already discussed.

The expression for the overall rate of photo-oxidation will then be the sum of the photo-oxidation rates of the accessible and partially inaccessible residue groups, and will be given by equation (2), if two different degrees of reactivity are involved

$$\frac{C}{C_O} \equiv f_1 e^{-k_1 t} + f_2 e^{-k_2 t} . \tag{2}$$

C/C_O is the fraction of the total amino acid residues, which is obtained from the results of the amino acid analyses carried out at various times after irradiation; f_1 and f_2 are the fractions of the total residues corresponding to the fast and slow reacting groups; and k_1 and k_2 are the first-order constants for the photo-oxidation of the respective residues. The values of f and k are obtained from a semilogarithmic kinetic plot like the one shown in Fig. 5.1.

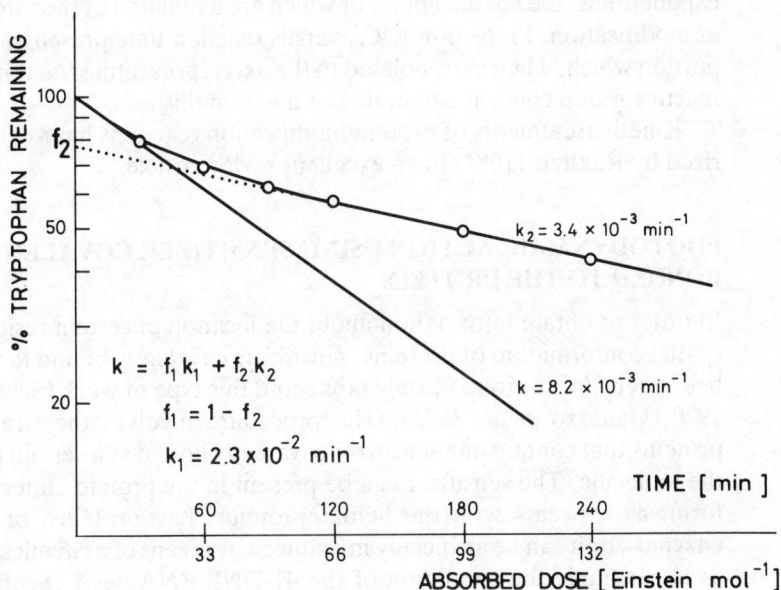

Fig. 5.1 — Tryptophan loss on photo-oxidation of α-lactalbumin in 0.05 M phosphate buffer pH 7.0 in the presence of riboflavin.

The situation described by equation (2) may be extended to cover cases with three or more sets of reactive groups.

In their original work, Ray & Koshland (1962) discussed in detail the advantages and limitations of their method. Among the advantages, it was stressed that no high specificity is required for the photo-oxidation, since the modification kinetics are analyzed in separate form for each amino acid residue. It is necessary, however, to point out that the presence of many residues of the same type can make the method more complex. In this way, the reaction of a single residue can be masked in the presence of many other residues of the same type that react with other reaction constants.

From the kinetic values obtained with this method, a correlation between the modification of amino acid residues and enzyme inactivation can be sought, which can lead to the determination of the number and type of residues that participate in the catalytic activity of an enzyme (Ray & Koshland 1962). For this type of correlation to be valid, it is necessary to make sure that during the photo-oxidation process no conformational changes take place. Photo-physical methods have been of great usefulness in the determination of conformational changes (Edwards & Silva 1985). In particular, the fluorescence techniques appear among the most convenient ones, because they are particularly sensitive to conformational changes Burstein *et al*. 1973).

In the cases of protein modification reactions where cooperativity is present, the reaction curves may be seen to consist of summations of exponentials, the coefficient (f) of which are a function of the rate constants of modification. Plots of ln C/C_O versus reaction time present a rectilinear portion which, when extrapolated to the axis representing fractional protein reactive group concentration, do not meet on this axis.

Kinetic treatments of protein modification reactions have been summarized by Rakitzis (1984) in an excellent review article.

PHOTODYNAMIC ACTION USING SENSITIZER COVALENTLY BONDED TO THE PROTEIN

In order to obtain information about the location of certain residues in the spatial conformation of proteins, sensitizers covalently bound to them have been used. Jori's group in Italy pioneered this type of work (Scoffone *et al*. 1970, Galiazzo *et al*. 1972). The procedure involves the irradiation of proteins that contain one sensitizer covalently bound at a certain position in the molecule. The sensitizer can be present in the protein either in natural form, as it occurs with the hematoproteins, flavoproteins, or pyridoxal enzymes, or it can be artificially introduced by means of a chemical reaction, as for example the formation of the 41-DNP-RNAase A (Scoffone *et al*. 1970). The selective modification of those residues that are located in a close proximity to the dye can be achieved by irradiation of the protein–sensitizer complex.

To get reliable results, some precautions must be taken when applying this methodology:

(a)   The insertion of the sensitizer into the molecule must bring about no significant alteration of the tertiary structure. A rigorous conformational analysis can be carried out, using the sensitive spectroscopic methods presently available (Jori *et al*. 1970, Burstein *et al*. 1973, Sun and Song 1977).

(b)   All photoreactions of intermolecular type must be avoided. With the purpose of minimizing this type of interaction, it is convenient to work with dilute solutions. In any case, the occurrence of intermolecular photosensitization can be easily detected by means of the irradiation of a mixture of labeled and unlabeled protein. If no intermolecular phenomena are produced, the unlabeled protein must be recovered without damage after irradiation.

(c) The binding of the sensitizer to the protein must be carried out with a high degree of selectivity, in such a way that the photodynamic action of the introduced sensitizer operates in a very well defined and restricted area.

REFERENCES

Amagasa, J. (1981) *Photochem. Photobiol.* **33** 947.

Asquith, R.S. & Rivett D.E. (1971) *Biochim. Biophys. Acta* **252** 111.

Bellin, J.S. (1968) *Photochem. Photobiol.* **8** 383.

Bellin, J.S. & Yankus, C.A. (1968) *Arch. Biochem. Biophys.* **123** 18.

Benassi, C.A., Scoffone, E., Galiazzo, G. & Jori, G. (1967) *Photochem. Photobiol.* **6** 857.

Burstein, E.A., Vedenkina, N.S. & Ivokova, M.N. (1973) *Photochem. Photobiol.* **18** 263.

Cadenas, E., Sies, H., Graf, H. & Ullrich, V. (1980) *Eur. J. Biochem.* **130** 117.

Calhoun, D.B., Vanderkooi, M.J. & Englander, S.W. (1983a) *Biochemistry* **22** 1533.

Calhoun, D.B., Vanderkooi, M.J., Woodrow, G.V. & Englander, S.W. (1983b) *Biochemistry* **22** 1526.

Creed, D. (1984a) *Photochem. Photobiol.* **39** 537.

Creed, D. (1984b) *Photochem. Photobiol.* **39** 563.

Dewey, D.L. & Stein, G. (1970) *Radiat. Res.* **44** 345.

Durán, N., Haun, M., De Toledo, S.M., Cilento, G. & Silva, E. (1983) *Photochem. Photobiol.* **37** 247.

Edwards, A.M. & Silva, E. (1985) *Radiat. Environ. Biophys.* **24** 141.

Eftink, M.R. & Jameson, D. (1982) *Biochemistry* **21** 4443.

Ferrer, I. & Silva, E. (1981) *Radiat. Environ. Biophys.* **20** 67.

Ferrer, I. & Silva, E. (1985) *Radiat. Environ. Biophys.* **24** 63.

Foote, C.S. (1968) *Acc. Chem. Res.* **1** 104.

Foote, C.S. (1976) In: *Free Radicals in Biology* (Pryor, W.A., ed.) p. 85, Academic Press, New York.

Fukunaga, Y., Katsuragi, Y., Izumi, T. & Sakiyama, F. (1982) *J. Biochem.* **92** 129.

Galiazzo, G., Jori, G. & Scoffone, E. (1968) *Biochem. Biophys. Res. Commun.* **31** 158.

Galiazzo, G., Jori, G. and Scoffone, E. (1972) In: *Research Progress in Organic Biological and Medicinal Chemistry III*, Part I, p. 137.

Ghiron, C.A. & Spikes, J.D. (1965) *Photochem. Photobiol.* **4** 13.

Grossweiner, L.I. (1969) *Photochem. Photobiol.* **10** 183.

Grossweiner, L.I. & Kepka, S.G. (1972) *Photochem. Photobiol.* **16** 305.

Grossweiner, L.I., Patel, A.S. & Grossweiner, J.B. (1982) *Photochem. Photobiol.* **36** 159.

Hasty, N., Merkel, P.B., Radlick, O. & Kearns, D.R. (1972) *Tetrahedron Lett.* **1** 49.

Hopkins, T.R. & Spikes, J.D. (1969) *Radiat. Res.* **37** 253.

Hopkins, T.R. & Spikes, J.D. (1970) *Photochem. Photobiol.* **12** 175.

Jori, G., Galiazzo, G., Marzotto, A. & Scoffone, E. (1968a) *J. Biol. Chem.* **243** 4272.

Jori, G., Galiazzo, G., Marzotto, A. & Scoffone, E. (1968b) *Biochim. Biophys. Acta* **154** 1.

Jori, G., Galiazzo, G. & Scoffone, E. (1969) *Int. J. Prot. Res.* **1** 289.

Jori, G. & Cauzzo, G. (1970) *Photochem. Photobiol.* **12** 231.

Jori, G., Galiazzo, G., Tamburro, M.A. & Scoffone, E. (1970) *J. Biol. Chem.* **245** 3375.

Jori, G. & Spikes, J.D. (1981) In: *Oxygen and Oxy-radicals in Chemistry and Biology* (Rodgers, M.A.J. & Powers, E.L., eds) p. 441, Academic Press, New York.

Kasha, A.U. and Branham, D.E. (1979) In: *Singlet Oxygen* (Wasserman, H.H. & Murray, R.W., eds) p. 1, Academic Press, New York.

Khan, A.U. & Kasha, M. (1970) *J. Am. Chem. Soc.* **92** 3293.

Khan, A.U. & Kasha M. (1979) *Proc. Natl. Acad. Sci. USA* **76** 6047.

Koizumi, M. & Usui, V. (1972) *Mol. Photochem.* **4** 57.

Kramer, H.F. & Maute, A. (1972) *Photochem. Photobiol.* **15** 7.

Lakowicz, J.R. & Weber, G. (1973) *Biochemistry* **12** 4171.

Linding, B.A. & Rodgers, M.A. (1979) *J. Phys. Chem.* **83** 1683.

Matsuura, T. (1977) *Tetrahedron* **33** 2869.

Merkel, P.B., Nilsson, R. & Kearns, D.R. (1972) *J. Am. Chem. Soc.* **94** 1030.

Nakagawa, M., Yojoyama, Y., Kato, S. & Hino, T. (1985) *Tetrahedron* **41** 2125.

Nilsson, R., Merkel, R.B. & Kearns, D.R. (1972) *Photochem. Photobiol.* **16** 117.

Nilsson, R. & Kearns, D.R. (1973) *Photochem. Photobiol.* **17** 65.

Ogilby, P.R. & Foote, C.S. (1983) *J. Am. Chem. Soc.* **105** 3423.

Ouannes, C. & Wilson, T. (1968) *J. Am. Chem. Soc.* **90** 6527.

Rakitzis, E.T. (1984) *Biochem. J.* **217** 341.

Ray, W.J., Jr. & Koshland, D.E., Jr. (1961) *J. Biol. Chem.* **236** 1973.

Ray, W.J., Jr. & Koshland, D.E., Jr. (1962) *J. Biol. Chem.* **237** 2493.

Reddi, E. Rodgers, M., Spikes, J. & Jori, G. (1984) *Photochem. Photobiol.* **40** 415.

Risi, S., Silva, E. & Dose, K. (1973) *Photochem. Photobiol.* **18** 475.

Rodgers, M.A. & Snowden P.T. (1982) *J. Am. Chem. Soc.* **104** 5541.

Rossi, E., Van de Vorst, A. & Jori, G. (1981) *Photochem. Photobiol.* **34** 447.

Saito, I., Yamane, M., Shimazu, N. & Matsuura, T. (1975) *Tetrahedron Lett.* **9** 641.

Saito, I., Matsugo, S. & Matsura, T. (1979) *J. Am. Chem. Soc.* **101** 4757.

Scoffone, E., Galiazzo, G. & Jori, G. (1970) *Biochem. Biophys. Res. Commun.* **38** 16.

Sconfienza, C., Van de Vorst, A. & Jori, G. (1980) *Photochem. Photobiol.* **31** 351.

Shugar, D. (1952) *Biochim. Biophys. Acta* **8** 302.

Silva, E. & Gaule, J. (1977) *Radiat. Environ. Biophys.* **14** 303.

Silva, E. (1979) *Radiat. Environ. Biophys.* **16** 71.

Silva, E., Risi, S. & Dose, K. (1974) *Radiat. Environ. Biophys.* **11** 111.

Sluyterman, L.A. (1962) *Biochim. Biophys. Acta* **60** 557.

Spikes, J.D. & Livingston, R. (1969) *Adv. Radiation Biol.* **3** 29.

Spikes, J.D. (1977) In: *The Science of Photobiology* (Smith, K.C., ed.) p. 87, Plenum Press, New York.

Sun, M. & Song, P.S. (1977) *Photochem. Photobiol.* **25** 3.

Tomita, M., Irie, M. & Ukita, T. (1969) *Biochemistry* **8** 5149.

Tsai, C.S., Godin, J.P.R. & Wand, A.J. (1985) *Biochem. J.* **225** 203.

Turro, N.J. (1978) *Modern Molecular Photochemistry*, Benjamin/Cummings, Menlo Park, California.

Weil, L., James, S. & Buchert, A.R. (1953) *Arch. Biochem. Biophys.* **46** 266.

Weil, L., Seibles.T.S. & Herskovits, T.T. (1965) *Arch. Biochem. Biophys.* **111** 308.

Zwicker, E.F. & Grossweiner, L.I. (1963) *J. Phys. Chem.* **67** 549.

6

The study of the catalytic sites of enzymes using fluorescent compounds

Dr Jorge E. Churchich, Department of Biochemistry, University of Tennessee, Knoxville, TN 37996, USA

INTRODUCTION

Fluorescence spectroscopy has proved to be a valuable technique in the study of ligand binding to enzymes. The method can provide information concerning the active sites of enzymes provided a fluorescent chromophore has reacted with specific amino acid residues located at the active center.

The fluorescence group should have absorption and fluorescence properties distinct from that of the tyrosyl and tryptophanyl residues of the protein.

In addition, it is desirable that the emission properties (fluorescence lifetime, emission maximum, and fluorescence quantum yield) are sensitive to the polarity of the environment of the chromophore.

In this chapter, the reactions of highly selective fluorescent probes with amino acid residues of enzymes are described.

ENZYMES INACTIVATED BY o-PHTHALALDEHYDE

o-Phthalaldehyde forms an isoindole adduct by reacting with the NH$_2$ group of an amino acid and β-mercaptoethanol. The structure of this adduct has been elucidated by Benson & Hare (1975).

o-Phthalaldehyde also reacts with proteins, in particular with cysteinyl and lysyl residues. The extent of the reaction with proteins is easily determined by measuring either the increase in absorbance at 335 nm or the increase in fluorescence at 450 nm. Free o-phthalaldehyde does not exhibit any fluorescence over the spectral range 400 to 500 nm upon excitation at 335 nm (Blaner & Churchich 1979).

The time-course of the reaction of succinic semialdehyde dehydrogenase with o-phthalaldehyde in the presence of 1 mM β-mercaptoethanol follows monophasic kinetics (Blaner & Churchich 1979). The binding of o-phthalaldehyde causes irreversible loss of catalytic activity, and experiments designed to correlate changes in enzymatic activity with the number of o-phthalaldehyde molecules reacted, revealed that complete inactivation of the dehydrogenase is achieved after the reaction of 2.5 moles of o-phthalaldehyde per subunit.

Palczewski *et al.* (1983) have characterized the inhibition of aldolase using o-phthalaldehyde. This compound forms an isoindole adduct by crosslinking ε-amino groups and sulfhydryl groups located at the catalytic site. The crosslinked adduct represented in Fig. 6.1 is fluorescent and

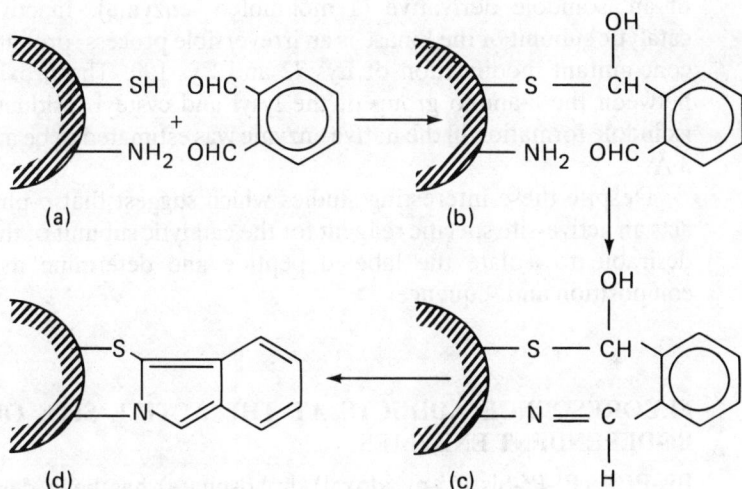

Fig. 6.1 — Formation of cross-linked adducts between o-phthalaldehyde and sulfhydryl and amino groups of enzymes.

facilitates the stochiometric determination of the number of modified active sites.

In order to gain information on the microenvironment of cysteinyl and lysyl residues participating in the formation of the isoindole derivative in the reaction between aldolase and o-phthalaldehyde, molar transition energies (E_T) of 1-(β-hydroxyethylthio)-2-β-hydroxyethyl-isoindole (also called EA adduct) in various solvents were compared with those of several isoindole derivatives. The molar transition energies were calculated from the fluorescence emission (λ em) of an isoindole using equation (1)

$$E_T = 2.985 \, \lambda \, \text{em} - 1087.28 \tag{1}$$

A value very close to 120 kJ/mole was obtained for the molar transition energy of the isoindole derivative formed between aldolase and o-phthalaldehyde; this value is comparable to that of EA adduct (synthetic isoindole) in hexane.

On the basis of these spectroscopic results, Palczewski *et al.* (1983) have concluded that the microenvironment of the active site of aldolase containing the cysteinyl and lysyl residues, participating in the reaction with o-phthalaldehyde, is fairly hydrophobic.

More recently, o-phthalaldehyde has been used as a probe of the active site of the catalytic subunit of adenosine cyclic-3'-5'-monophosphate dependent protein kinase from bovine skeletal muscle (Puri *et al.* 1985). Absorbance and fluorescence spectroscopy data were consistent with the formation of an isoindole derivative (1 mol/mol of enzyme). Inactivation of the catalytic subunit of the kinase is an irreversible process, probably due to the concomitant modification of Lys-72 and Cys-199. The proximal distance between the ε-amino group of the lysyl and cysteyl residues involved in isoindole formation in the native enzyme was estimated to be approximately 3 Å.

Despite these interesting studies which suggest that o-phthalaldehyde acts an active-site specific reagent for the catalytic subunit of the kinase, it is desirable to isolate the labeled peptide and determine its amino acid composition and sequence.

FLUORESCENCE ADDUCTS AT THE ACTIVE SITE OF VITAMIN B6-DEPENDENT ENZYMES

Bis-PLP (P^1-P^2-bis (5'-pyridoxal)-diphosphate) has been described as an active site specific reagent of phosphorylase (Shimomura & Fukui 1978). The bifunctional reagent binds to the PLP site in apo-phosphorylase resulting in the formation of a catalytically inactive species. Since bis-PLP is a specific reagent for ε-amino groups of lysyl residues, it was not surprising to find that it reacts with lysyl residues other than those involved in the binding of the cofactor pyridoxal-5-P.

The effect of bis-PLP on the catalytic function of 4-aminobutyrate aminotransferase was investigated by preincubating the enzyme with increasing concentrations of bis-PLP at 25°C (Kim & Churchich 1982). Inactivation was attained when the aminotransferase was preincubated with 20-fold molar excess of bis-PLP at pH 7 for 60 minutes. Although the catalytic activity could not be restored by addition of PLP, complete protection against the inhibitory effect of bis-PLP was observed in the presence of α-ketoglutarate. As shown in Fig. 6.2, the time-course of inactivation of 4-aminobutyrate aminotransferase is significantly affected by variations in the pH of the reaction mixtures.

If the reaction of the lysyl group of the protein with bis-PLP can be described by the simple reaction included in the scheme

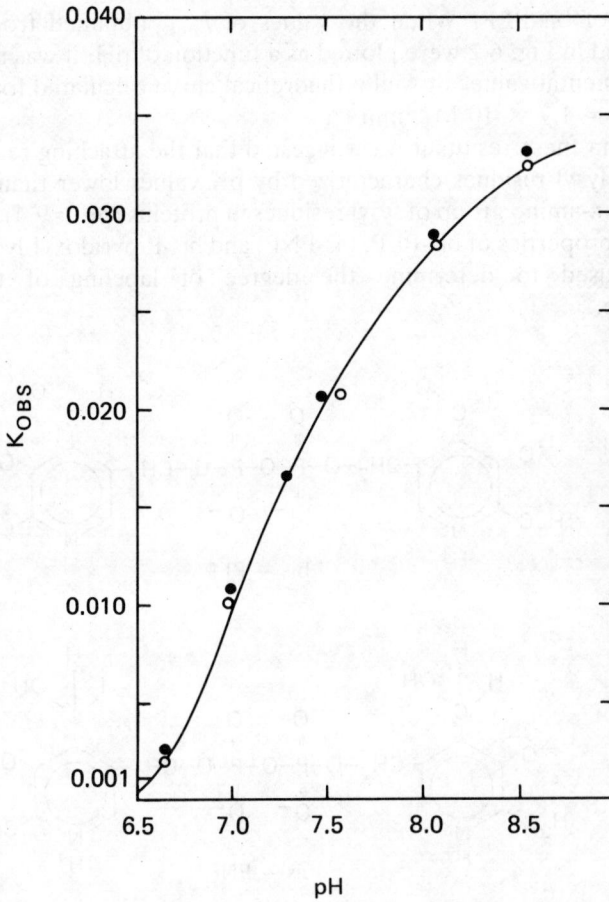

Fig. 6.2 — Plot of k_{obs} (min^{-1}) versus pH. The observed rate constant of inactivation (k_{obs}) (o) was determined from the results included in Fig. 6.3. The theoretical values (●) were calculated for pK = 7.3 and $k_0 = 1.7 \times 10^2$ M^{-1} min^{-1}.

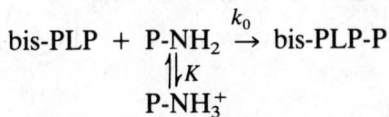

$$\text{bis-PLP} + \text{P-NH}_2 \xrightarrow{k_0} \text{bis-PLP-P}$$
$$\updownarrow K$$
$$\text{P-NH}_3^+$$

then, the observed rate constant of inactivation (k_{obs}) is related to the dissociation constant (K) and the proton concentration (H$^+$) by equation 2

$$k_{obs} = \frac{k_0 (\text{PLP})_2}{1 + \text{H}^+/K} \tag{2}$$

where k_0 is the second -order rate constant and (PLP)$_2$ the initial concentration of bis-PLP. When the values of k_{obs}, obtained from the results included in Fig. 6.2 were plotted as a function of pH, it was found that the experimental values fit well a theoretical curve calculated for $K = 10^{-7.3}$ and $k_0 = 1.7 \times 10^2 M^{-1} min^{-1}$.

From these results it was suggested that the attacking reagent bis-PLP blocks lysyl residues characterized by pK values lower than those of the normal ε-amino group of lysyl residues in proteins (pK = 9.4). The spectroscopic properties of bis-PLP, bis-PNP, and bis-P-pyridoxyl-lysine (Fig. 6.3) were used to determine the degree of labeling of the modified enzyme.

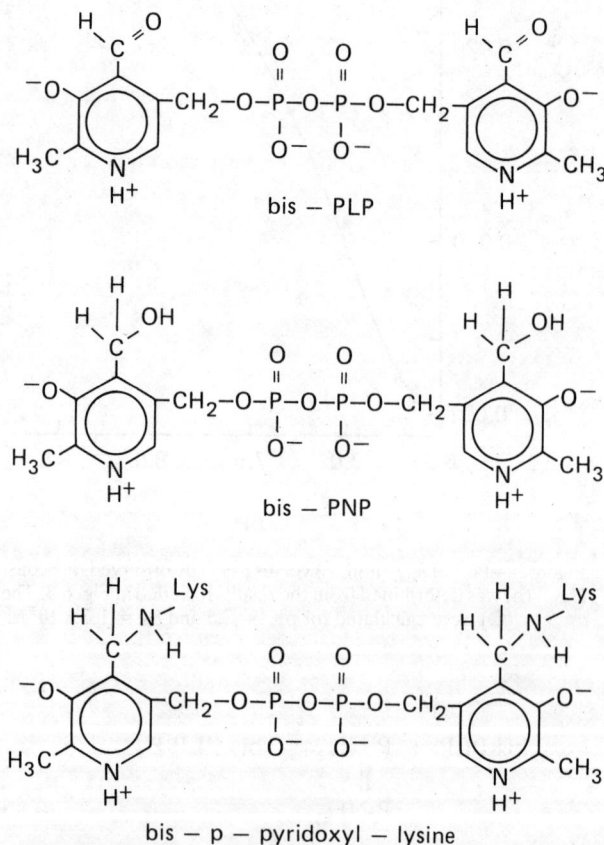

Fig. 6.3 — Structure of P1,P2-bis (5′ pyridoxal) diphosphate (bis-PLP), P1,P2-bis (5′-pyridoxine) diphosphate (bis-PNP), and bis-P-pyridoxyllysine.

The reaction of bis-PLP with ε-amino groups of lysyl residues leading to the formation of Schiff's base adducts is accompanied by characteristic changes in the absorption spectrum of the modifier; the absorbance at 391 nm decreases with a concomitant increase in absorbance at wavelengths longer than 400 nm. Upon reduction with $NaBH_4$, the modified enzyme reacted with bis-PLP displays an intense absorption band at 325 nm, together with an intense fluorescence band centered at 395 nm.

The intense absorbtion band at 325 nm can be attributed either to pyridoxine-5-P or P-pyridoxyl residues. Like pyridoxine-5-P, bis-pyridoxine-5-P interacts with borate to form a stable adduct which absorbs maximally at 290 nm. P-pyridoxyl-lysine does not form a complex with borate and the absorption band at 325 nm remains essentially invariant upon addition of borate. Hence, the degree of labeling of the aminotransferase reacted with bis-PLP and reduced with $NaBH_4$ can be determined by absorption spectroscopy of samples in the absence and presence of borate.

If, on the other hand, a fraction of the carbonyl groups of the bifunctional reagent has not reacted with the protein, then the absorption band at 325 nm should be perturbed by borate ions. When this spectroscopic method was applied to the determination of the number of P-pyridoxyl residues introduced into the aminotransferase, it was observed that 2.1 P-pyridoxyl and 2.2 P-pyridoxine residues were present per enzyme dimer.

The ratio of P-pyridoxyl to P-pyridoxine residues is, within experimental error, equal to one, indicating that the bifunctional reagent, whose expanded structure covers a distance of the order of 17 Å, has failed to crosslink lysyl residues located on different subunits. Further support for this contention was derived from electrophoresis experiments conducted on sodium dodecyl-sulfate polyacrylamide gels (Kim & Churchich 1982).

REACTION OF GABACULINE WITH VITAMIN B_6 DEPENDENT ENZYMES

L-gabaculine (5-amino-1,3-cyclohexane dienylcarboxylate), a neurotoxin first isolated from Streptomyces toyocaensis by Kobayashi *et al.* (1977) is a potent inhibitor of both mammalian and bacterial 4-aminobutyrate-α-ketoglutarate-aminotransferases. The inactivation of ω aminotransferases was shown to proceed via a pathway termed 'aromatization' (Rando & Bangerter 1976) (Fig. 6.4), in which the initial Schiff's base between

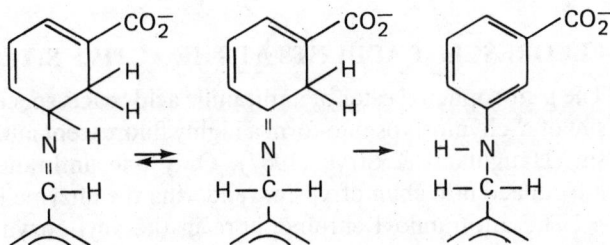

Fig 6.4 — Aromatization pathway for enzyme inhibition by L-gabaculine.

gabaculine and the cofactor pyridoxal-5-P is converted into a stable secondary amine, *m*-carboxyphenyl-pyridoxamine-5-P. The secondary amine remains firmly bound to the catalytic site of the aminotransferase, but can be removed either by extensive dialysis or by addition of pyridoxine-5P oxidase which catalyzes the oxidation of *m*-carboxyphenyl-pyridoxamine-5-P to pyridoxal-5-P (Kim *et al.* 1981).

Dissociation of the modified transaminase into free apoenzyme and *m*-carboxyphenyl-pyridoxamine-5-P is easily monitored by fluorescence spectroscopy, since the fluorescence of the bound PLP-adduct is quenched when compared to free *m*-carboxyphenyl-pyridoxamine-5-P.

L-gabaculine is structurally related to 4-aminobutyrate, and, therefore, its spectrum of action was originally considered to be exclusive for enzymes whose catalytic specificity was related to the turnover of 4-aminobutyrate. However, it was demonstrated by Soper & Manning (1982) that enzymes for which 4-aminobutyrate is not a substrate are also inactivated by gabaculine.

The effects of the inhibitor on such diverse groups of enzymes, i.e. D-amino acid aminotransferase, L-alanine aminotransferase, and L-aspartate aminotransferase, appear to be related to the enzymic exchange of β-protons of their normal substrates.

As indicated in Fig. 6.4, the aromatization of gabaculine requires two successive proton translocations; the first proton to be removed is on the carbon atom adjacent to the amino group. It is well established that pyridoxal-5-P dependent enzymes remove this proton from their normal substrates during catalysis.

The next step in the aromatization of gabaculine requires the removal of a second proton on the carbon atom adjacent to the first carbon, a step analogous to the removal of protons from the β-carbon of normal substrates. Enzymes such as L-alanine aminotransferase and aspartate aminotransferase, which catalyze the exchange of β-protons from their substrates, are inactivated by gabaculine, whereas alanine racemase and tryptophanase, which do not exchange β-protons from their substrates, are resistant to inactivation by gabaculine. All these observations are consistent with the hypothesis that aromatization of gabaculine takes place when the enzyme is capable of exchanging β-protons from their normal substrates.

FLUORESCENT ADDUCTS AT THE ACTIVE SITES OF PROTEASES

The *p*-nitrophenyl ester of anthranilic acid reacts specifically with the active site of α-chymotrypsin to form a highly fluorescent anthraniloyl chymotrypsin (Haughland & Stryer 1967). Only one anthraniloyl chromophore is introduced into chymotrypsin, rendering the enzyme inactive.

The anthraniloyl chromophore in the acyl enzyme can be selectively excited since its absorption (342 nm) and emission (422 nm) maxima are distinct from those of the aromatic residues of proteins. Using steady and nanosecond fluorescence spectroscopy, Haughland & Stryer (1967) determined the flexibility of the chromophore bound to a seryl residue at the active site of chymotrypsin. The *rotational correlation* time of the conjugate protein was found to be 16 ns, indicating that the active site of the modified enzyme is rather rigid.

Although blocking of seryl residues in either chymotrypsin or trypsin leads to loss of catalytic activity, the modified proteins can be used as 'probes' of protein–protein interactions. In our laboratory, we have investigated the spectroscopic properties of anthraniloyl-trypsin when this modified protein is allowed to interact with human α_2-macroglobulin (Churchich, unpublished results). Human α_2-macroglobulin is a tetrameric plasma glycoprotein that acts as an inhibitor of a wide variety of proteinases (Barret & Starkey 1973, Harpel 1973).

The interaction between trypsin and α_2-macroglobulin is initiated by binding of the proteolytic enzyme to the so called 'bait region' located near the middle of a polypeptide chain of molecular weight 180 000 (Sottrup *et al.* 1980). Restricted proteolysis takes place, and four thiol groups are generated per mol of α_2-macroglobulin (Van Leuven 1982).

Chemical and structural events following the limited proteolysis of α_2-macroglobulin have been carefully examined in several laboratories (Bieth *et al.* 1981), but there is little information related to the primary event itself, i.e. the binding of trypsin to α_2-macroglobulin prior to proteolysis. Does the α_2-macroglobulin recognize the proteolytic enzyme even when its catalytic site is blocked by an anthraniloyl chromophore?

Anthraniloyl trypsin binds to α_2-macroglobulin as demonstrated by the increase in emission anisotropy values upon addition of increasing concentrations of modified trypsin to a fixed concentration of macroglobulin (Fig. 6.5). Dilution of the reaction mixture to a final concentration of 0.6 μM α_2-macroglobulin has no effect on the emission anisotropy when the molar mixing ratio of anthraniloyl-trypsin to α_2-macroglobulin is one to one.

If trypsin labeled with the fluorescence probe is trapped by α_2-macroglobulin, then one should observe a decrease in the accessibility of the anthraniloyl chromophore to collisional encounters with quencher molecules. In the event of collisional quenching, the ratio of fluorescent lifetimes is equal to the ratio of fluorescence yields, and the plot of F_0/F versus quencher concentration (Q) would yield a straight line whose slope is equal to the Stern–Volmer's constant (K_{sv}). When free and bound anthraniloyl-trypsin were examined under conditions of KI quenching, the collisional rate constant (K_q) for free anthraniloyl-trypsin was found to be at least five-fold greater than the collisional rate constant determined for complexed anthraniloyl-trypsin.

The steady-state measurements already indicate interaction between catalytically inactive trypsin and α_2-macroglobulin. However, it was desirable to determine the degree of rotational mobility of trapped anthraniloyl-trypsin by measuring rotational correlation times in the absence and presence of α_2-macroglobulin. To this end the technique of time-correlated single proton counting was applied to the determination of rotational correlation time values (O'Connor & Phillips 1984). Fig. 6.6 shows the results of the time-dependent anisotropy measurements, about which the

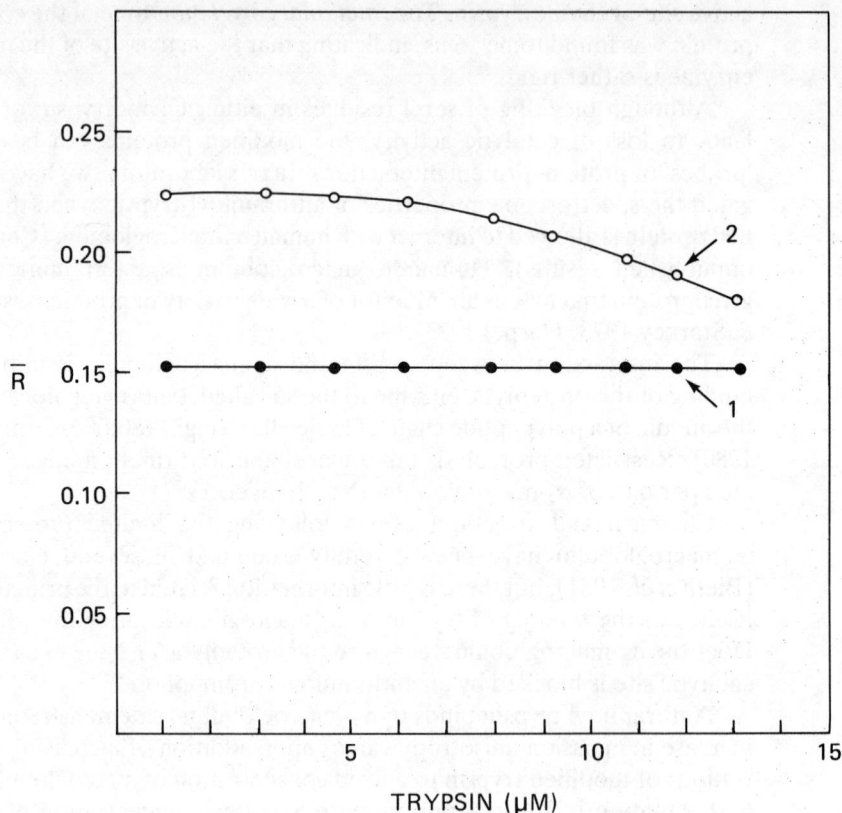

Fig. 6.5 — Steady emission anisotropy values of anthraniloyl-trypsin (1) and α_2-macroglobulin (4μM) in the presence of increasing concentrations of anthrani-loyl-trypsin (2). Excitation: 345 nm, emission: 460 nm. Results obtained at 20°C.

following points are noted: (a) the emission anisotropy decay of free anthraniloyl-trypsin is accurately described by the equation

$$R(t) = R(0) \, e^{-t/\phi} \tag{3}$$

(b) Anthraniloyl-trypsin in the presence or α_2-macroglobulin exhibits bi-exponential kinetic decay. Two rotational correlation times $\phi_1 = 8$ ns and $\phi_2 = 132$ ns are obtained for $\alpha_1 = 0.48$ and $\alpha_2 = 0.62$. The results fit an equation of the form

$$R(t) = R(0) \, \alpha_1 e^{-t/\phi_1} + R(0) \, \alpha_2 e^{-t/\phi_2} \tag{4}$$

The nature of the processes leading to multi exponential decay of the emission anisotropy function deserve attention and it may be discussed in reference to a model proposed by Kinosita *et al.* (1977).

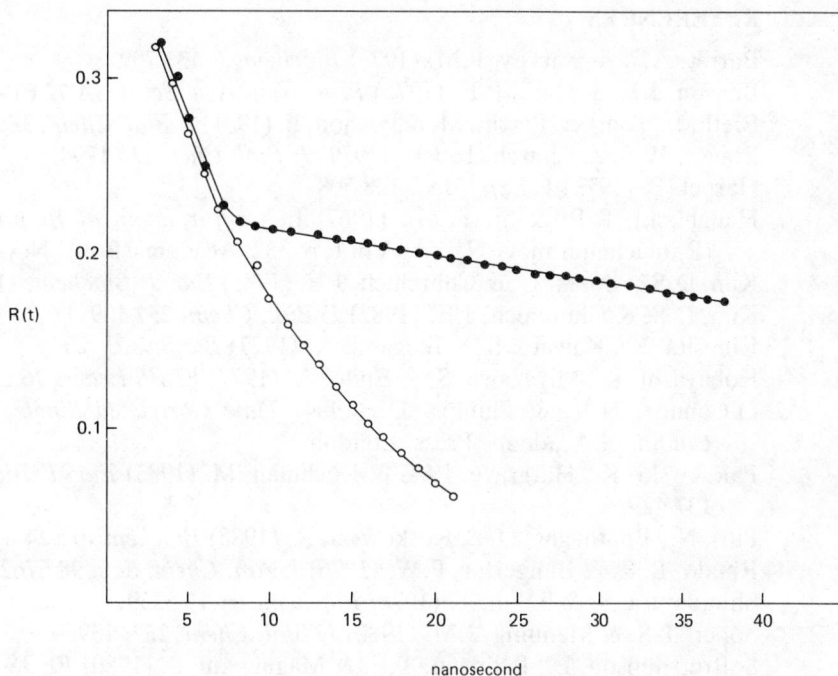

Fig. 6.6 — Time emission anisotropy decay of anthraniloyl-trypsin (○) and anthraniloyl-trypsin (2 μM) + α2-macroglobulin (4 μM) (●). Excitation: 345 nm, emission:
460 nm.

As shown in Fig. 6.6, the rotational motion of antraniloyl-trypsin is
hindered by the presence of α_2-macroglobulin, but the emission anisotropy
function $R(t)$ does not decay to zero. Indeed, a limiting anisotropy value
(R_∞) is observed at times which are long compared to the fluorescence
lifetime ($\tau = 7.2$ ns). The rapid localized motion of anthraniloyl-trypsin is
confined to the surface of a cone of semi-angle (A), which is related to the
ratio R_∞/R_0.

$$R_\infty/R_0 = [\tfrac{1}{2}(\cos A) \cdot (1 + \cos A)]^2 \tag{5}$$

For $R_\infty = 0.18$ and $R_0 = 0.36$, an angle $A = 48°$ is obtained. This angle
represents the semi-angle of a cone in which anthraniloyl-trypsin is free to
wobble. The same experimental approach could be used to study the
behavior of enzymes specifically tagged with fluorescent probes. Time-
resolved emission anisotropy is one of the few techniques that provide
reliable information on fluctuations of protein structure in the nanosecond
range.

ACKNOWLEDGEMENT

This work was supported by NIH grant GM 27639-04.

REFERENCES

Barret, A.J. & Starkey, P.M. (1973) *Biochem. J.* **133** 709.

Benson, J.R. & Hare, P.E. (1975) *Proc. Natl. Aca. Sci. USA* **72** 619.

Bieth, J., Tourbez-Perrin, M, & Pochon, F. (1981) *J. Biol. Chem.* **256** 7954.

Blaner, W.S. & Churchich, J.E. (1979) *J. Biol. Chem.* **24** 1794.

Harpel, P. (1973) *J. Exp. Med.* **138** 508.

Haughland, R.P. & Stryer, L. (1967) In *Conformation of Biopolymers* (Ramachandran, G.N., ed.) Vol 1, p. 332, Academic Press, New York.

Kim, D.S., Moses, U. & Churchich, J.E. (1981) *Eur. J. Biochem.* **118** 303.

Kim, D.S. & Churchich, J.E. (1982) *J. Biol. Chem.* **257** 10991.

Kinosita, K., Kawato, S. & Ikegami, A. (1977) *Biophys. J.* **20** 289.

Kobayashi, K., Miyazawa, S. & Endo, A. (1977) *FEBS Letters* **76** 207.

O'Connor, D.V. & Phillips, D. (1984) *Time Correlated Single Proton Counting*, Academic Press, London.

Palczewski, K., Hargrave, P.A. & Kochman, M. (1983) *Eur. J. Biochem.* **137** 429.

Puri, N., Bhatnagar, D. & Roskowski, R. (1985) *Biochemistry* **24** 6499.

Rando, R.R. & Bangerter, F.W. (1976) *J. Am. Chem. Soc.* **98** 6762.

Shimomura, S. & Fukui, T. (1978) *Biochemistry* **17** 5359.

Soper, T.S. & Manning, J.M. (1982) *J. Biol. Chem.* **257,** 13930.

Sottrup-Jenson, L., Petersen, T.E. & Magnusson, S. (1980) *FEBS Letters* **121,** 275.

Van Leuven, F. (1982) *Trends Biochem. Sci.* **7** 18.

7

Chemical modification of allosteric properties

Dr Robert G. Kemp, Department of Biological Chemistry &
Structure, The University of Health Sciences/The Chicago Medical
School, North Chicago, IL 60064, USA

As indicated in other chapters of this book, chemical modification of
enzymes can be monitored by changes in activity of the enzyme. Conclusions
regarding the nature of the modification can be derived by correlating these
activity changes with data on the incorporation of the modifying reagent,
which is usually facilitated by the presence of a chromophore or radiolabel
on the reagent. The ideal situation is regarded as one in which there is a
direct stoichiometric relationship between enzyme activity and the incorporation of reagent into a single amino acid residue, either one involved in
substrate binding or directly in the catalytic process. Example of these are
the classic studies of the modification of pancreatic proteases with diisopropylfluorophosphate or with chloromethylketone affinity reagents. Chemical
modification of allosteric enzymes, on the other hand, provides a great deal
more information while providing additional problems of monitoring the
modification event and interpreting the data that are obtained. For example,
site occupancy in a non-allosteric enzyme necessarily produces loss of
activity, while occupancy of an allosteric site can produce several effects.
Instead of monitoring activity loss, one is forced to measure loss of binding
capacity to provide additional evidence for site occupancy by a modifying
chemical reagent.

In addition to providing evidence concerning the make-up of the catalytic site, chemical modification of allosteric enzymes can provide information about critical residues in an allosteric site, provide a means of
studying an enzyme that has been desensitized to allosteric control, establish

that a regulatory site is topographically distinct from the catalytic site, and produce an enzyme that has been frozen into one of multiple conformation states available to the native enzyme. In addition, once a procedure has been established for the production of an enzyme modified at a particular site, reagents bearing spin labels or fluorescent probes can be utilized to map inter-site distances within the molecule. In addition, hybrid molecules can be produced containing native and modifed subunits to provide information concerning the transmission of conformational events and intersubunit communication.

As indicated, there are a number of potential consequences of chemical modification of allosteric enzymes. Activity losses can occur as a result of a specific modification of critical residue or residues in the substrate binding pocket or of those residues intimately involved in the catalytic process. Generalized loss of activity can also occur as a result of modifications that produce discrete conformational changes that disrupt the active-site or more generalized disruption of overall secondary, tertiary, and/or quaternary structure of the protein. Of greater interest here are those modifications that result in alterations of regulatory properties. Such modifications may be grouped into three general types. First, the modification can be a productive interaction with the regulatory site in that the enzyme behaves as it does when the site is occupied by the natural dissociable ligand. In this instance the enzyme is frozen irreversibly in a particular conformation, with higher or lower activity, depending upon the natural ligand's action. Second, the modification can react with residue(s) within the allosteric site to prevent interaction of the natural ligand while not yielding a productive conformational change on its own. This would desensitize the enzyme at that particuar site. It should be noted that partial effects can be produced. Because a given binding pocket contains a number of residues important to binding, destruction of any one may simply decrease, but not abolish, affinity for the regulatory ligand. The third class of modification is represented by those that influence the allosteric transitions or conformational changes that occur as a result of interactions with regulatory ligands. In other words, binding of regulatory ligand occurs, but no regulation results. The most readily observed of this last type are those that result in dissociation of multi subunit enzymes, thus preventing inter-subunit transmission of conformational events associated with substrate or regulatory ligand interaction. One can also conceive of modifications that block a residue critical to intra- or inter-subunit communication that would result in some degree of loss in the transmission of conformational events.

In this chapter, the aforementioned consequences of chemical modification of allosteric enzymes will be discussed with concrete examples. In addition, the methodology for studying those changes that occur upon chemical modification will be presented. Most of the discussion will relate to work done by the author and others with an enzyme of exquisite allosteric complexity: mammalian phosphofructokinase. The following section will briefly describe the properties of this enzyme.

PHOSPHOFRUCTOKINASE: A MODEL OF ALLOSTERISM

Mammalian phosphofructokinase, which carries out the phosphorylation of fructose 6-P with ATP to produce fructose 1,6-P_2 and ADP, is tetrameric with a subunit size of approximately 80 kd. The enzyme is under regulatory control and is thought to be the principal pacemaker in glycolysis. The most important regulatory properties are typified in the three panels of Fig. 7.1.

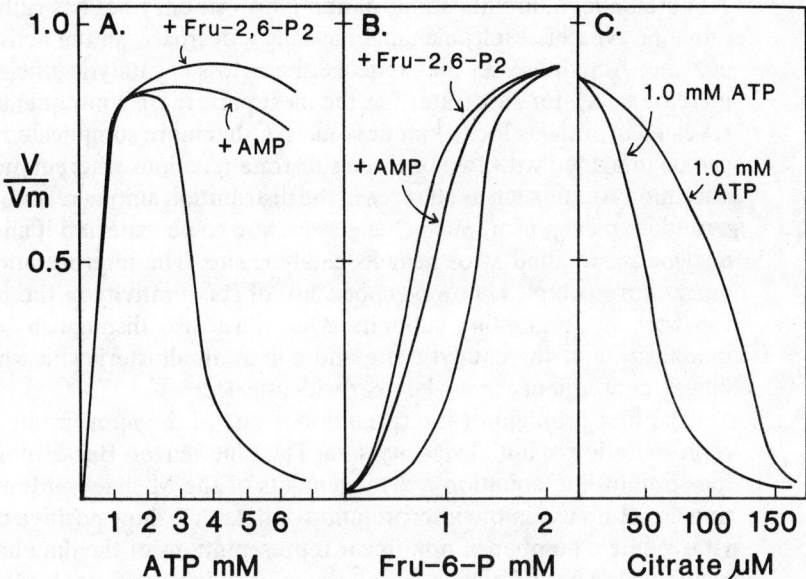

Fig. 7.1 — Allosteric properties of phosphofructokinase. Panel A, typical ATP inhibition curve at pH 7.2 plus and minus 20 μM AMP or 0.5 μM fructose-2,6-P_2. Panel B, fructose-6-P saturation curve ± AMP or fructose-2,6-P_2. Panel C, citrate inhibition at two non-inhibitory concentrations of ATP. Note synergism.

The enzyme is inhibited by high concentrations of one of the substrates, ATP (Fig. 7.1A); and the inhibition is reversed in the presence of the activators AMP (or ADP) and fructose 2,6-P_2 (or fructose 1,6-P_2). The enzyme displays a striking sigmoid response toward the concentration of the other substrate, fructose 6-P (Fig. 7.1B); and the activators produce a decrease in fructose 6-$P_{0.5}$ and a decrease in the interaction coefficient (slope in a Hill-type plot). The action of the other major inhibitor citrate is synergistic with ATP inhibition (Fig. 7.1C). Binding studies have shown that each subunit has six organic ligand binding sites: MgATP and fructose 6-P (the catalytic site), MgATP and citrate at inhibitory sites, and sites for each of the activators, AMP and fructose bisP$_2$. Inorganic phosphate and ammonium ion are also activators, but the number of sites of interaction has not been determined. For a more extensive discussion of these and other characteristics of phosphofructokinase, the reader is recommended to reviews by Uyeda (1979) and Kemp & Foe (1983).

METHODS TO STUDY THE CONSEQUENCES OF MODIFICATION OF ALLOSTERIC ENZYMES

Kinetic studies

As noted above, modifications of non-allosteric enzymes basically produce only one type of result, a change, usually a decrease, in the activity of the enzyme. Activity losses may reflect either a loss in catalytic efficiency or an increase in K_m for substrate. For the most part, these consequences will be revealed in initial velocity kinetic studies. Other more complicated scenarios can be imagined with two or three substrate reactions wherein mechanistic alterations occur such as changes in the distribution among reaction paths in a random mechanism. Such changes are also to be expected if an allosteric enzyme is modified at or near its catalytic site. The interpretation here is much more complex, however, because of cooperativity in the binding of substrate by interacting subunits. One must also distinguish between a modification at the catalytic site and one at an allosteric site wherein the kinetic consequences may be nearly identical.

The first problem of interpretation is one of the appropriate graphical representation of initial velocity data. The Lineweaver–Burke plot or any of the straight-line equation rearrangements of the Michaelis–Menten equation are of no use in the interpretation of data that show positive cooperativity. While a number of non-linear representations of the data have utility including the basic V versus S plot, the most widely used graphical representation is the Hill plot for the binding of oxygen to hemoglobin as adapted to kinetic data. Usually called the 'Hill-type' plot, it plots the log $V/(V_{max} - v)$ versus log S, and gives useful straight-line data between velocity values of 10 to 90% of V_{max}. The slope of the Hill plot yields a number usually referred to as the 'interaction coefficient' (or simply n), which is a measure of the degree of cooperativity in the binding of substrate or ligands. An enzyme that obeys Michaelis–Menten kinetics would give an interaction coefficient of one. Positive cooperativity gives values greater than one, negative cooperativity yields values less than one. An example of the use of the Hill-type plot will be presented later in this chapter (see Fig. 7.5).

Ligand binding studies

Kinetic studies of allosteric enzymes often cannot distinguish between a modification directly at a substrate site versus one that occurs at an allosteric site and influences the catalytic site. Similarly, one cannot distinguish kinetically from the loss of interaction at a regulatory site versus a disruption in conformational signal, such as that produced by dissociation of subunits. Ligand binding studies provide information about the site that has undergone modification. One compares native and modified enzyme in terms of both the maximum number of moles of ligand bound as well as the affinity of the enzyme for the ligand. This requires that binding be performed at varying substrate concentration and the data plotted appropriately. If the binding adheres to a hyperbolic isotherm, then any of the familiar graphical techniques of enzyme kinetics can be employed, substituting moles bound for the velocity term. The most familiar graphical representation is that of the Scatchard-type plot, wherein C, the number of moles bound per mole of protomer, divided by L, the concentration of free ligand, is plotted against C. The slope provides the dissociation constant while the X intercept gives the binding unit. The Scatchard plot often gives interpretable data for negative cooperativity, half-the-sites reactivity, or multiple binding sites.

One of the earliest binding methods employed was that of equilibrium dialysis (e.g. Changeux *et al.* 1968). Ligand and protein are mixed and placed in a dialysis bag suspended in an appropriate buffer. At equilibrium the concentration of ligand outside the bag represents free ligand, and that inside is bound plus free. The method has the disadvantage of being slow because of the time required to achieve equilibrium. An interesting variation on the theme of dialysis was the approach of Colowick & Womack (1969) who devised an analysis of the free ligand concentration based upon the rate of flow across a semipermeable membrane into a chamber that was to be continuously flushed into a fraction collector. In such a system a single protein sample can be used for an entire binding curve, and a whole series of data points can be collected within an hour. The technique works most effectively with radiolabeled ligands available at a fairly high specific activity and when the affinity is high. A precise but tedious variation on equilibrium dialysis is the method of Hummel & Dreyer (1962). The protein is passed through a column of Sephadex G25 or G50 previously equilibrated with ligand, usually radiolabeled. The protein will continue to bind more ligand until it comes into equilibrium with the concentration of ligand on the column. Fractions are collected, and the protein peak will be the sum of free and bound ligand, while the plateaus before and after the peak represent free ligand. The shape of the elution profile provides clues to non-equilibrium or to other artifacts such as instability of the protein or the ligand. A column can be run in an hour; but when one includes the work of collecting and analyzing fractions, the method is slow.

Another technique that uses the exclusion properties of Sephadex to separate free ligand solution from protein plus ligand is a method described by Kitajima & Uyeda (1983). Small columns of Sephadex are placed in a centrifuge tube and sedimented at low speeds to facilitate the separation.

Another method is the membrane filtration technique which has been described by Paulus (1969). Free ligand concentration is determined from a filtrate after pushing a portion of ligand–protein mixture through an ultra-filter. This is a fairly widely used technique, although problems can occur if either the ligand or the protein interact with the filter.

Several of these techniques find their antecedent in the sedimentation technique that was apparently first utilized by Hayes & Velick (1954).

Binding of NAD to a dehydrogenase was demonstrated by centrifuging protein plus NAD at high speeds to concentrate the protein at the bottom of the tube. Samples in the protein-free supernatant and the protein layer could be analyzed for NAD to determine free and bound plus free, respectively. While this technique was time-consuming as first described, the more recent utilization of ultracentrifuges that handle very large numbers of samples makes the technique more attractive. The problem with this and any technique that leads to concentration of the protein is that misleading results will be obtained if the protein shows concentration-dependent changes in ligand-binding behavior. Such behavior is not unusual among allosteric enzymes.

An entirely different class of binding techniques involves the use of optical methods. Changes in fluorescence of the ligand upon binding can be readily quantitated and saturation curves developed. This method has been widely used to study the binding of NAD and its analogues to dehydrogenases. Fluorescent derivatives of ligands that do not normally display fluorescence have been used, such as the etheno-derivatives of adenine nucleotides (e.g., Liou & Anderson 1978). Still another binding technique follows the quenching of intrinsic fluorescence of tyrosine residues in the protein. One note of caution here, however, is that the fluorescence change can result from a conformational event associated with binding. Because a modification could disrupt the sequence of events that produces that conformational change without greatly affecting the binding event itself, a failure to elicit a change in intrinsic fluorescence typical of the native enzyme upon exposure to ligand does not necessarily imply loss of binding.

Conformational studies

It is possible that binding may not elicit the conformational change required to change activity. The classic example of this are the studies of aspartic transcarbamylase where binding of regulatory ligand is not decreased despite the loss of an allosteric effect by dissociation of the catalytic and regulatory subunits (Changeux *et al.* 1968). Furthermore, occupancy of the same site by two different ligands does not necessarily produce the same conformational or kinetic result (Kemp & Foe 1983). Obviously, a study of the consequences of chemical modification should include evidence for, or excluding, perturbation in the normal pattern of conformational change associated with the transition between activated and inhibited state. Changes in protein conformation can be determined by a variety of optical techniques that have been employed for many years to study ordered structures in proteins. These include optical rotatory dispersion, circular dichroism, and intrinsic fluorescence measurements. Changes in polymeric state can be monitored by analytical ultracentrifugation, by analysis of Stokes radius by gel exclusion chomatography, or by sucrose gradient sedimentation. Differential reactivity of functional residues on the protein surface also provides evidence for different protein conformations under varying conditions. Trace labeling of amino groups with acetic anhydride or other highly reactive acylating agents coupled with HPLC mapping of tryptic digests can provide evidence for different conformational states. The author's laboratory has used extensively the reactivity of thiol groups as a conformational monitor. Thiol groups are attractive because they can be readily modified by a great number of reagents such as a variety of disulfide interchange reagents (e.g. Ellman's reagent) and a number of alkylhalides. Still another indicator of conformational events is differential reactivity toward proteolytic enzymes. In any of the techniques based on differential reactivity there remains one uncertainty: a loss in reactivity of a protein side-chain or the diminished susceptibility of a particular peptide bond to a protease can result either from a direct shielding of the residue or protease recognition site by a ligand or from a conformational event associated with ligand binding. Other evidence based on results from other ligands or on other types of conformational analysis would be required to distinguish between these possibilities.

STUDIES ON THE MODIFICATION OF PHOSPHOFRUCTOKINASE

Modification studies of phosphofructokinase have been extensive, leading to all of the potential consequences mentioned previously, that is, productive modifications to freeze the enzyme in activated or inhibited conformations, modifications that prevent ligand induced action by blocking a regulatory site, and modifications that interfere with conformational transitions. An example of the first type is shown in studies by Mansour & Colman (1978) who employed the adenine nucleotide analogue, F-sulfonylbenzoyladenosine. These studies originally performed on sheep heart muscle phosphofructokinase produced an activated enzyme largely desensitized to ATP inhibition. We have repeated this modification on rabbit skeletal muscle enzyme under the identical conditions described by Mansour & Colman. The effects of modification on enzyme activity is shown in Fig. 7.2A. The modified enzyme is much less sensitive to ATP inhibition, although very high concentrations do inhibit (not shown). That the AMP binding site is blocked is shown by the failure of the modified enzyme to bind cyclic AMP as indicated by the Hummel–Dreyer binding column profiles (Fig. 7.2B). The upper curve describes the elution of native enzyme that has bound [^{3}H]cyclic AMP while passing through a column equilibrated with 1 μM labeled cyclic AMP. The lower part of the figure shows that the modified enzyme does not bind cyclic AMP, presumably because the binding site is occupied by the covalently-bound nucleotide analogue.

A covalent modification at an inhibitory site is typified by the modification of phosphofructokinase by pyridoxal phosphate (Colombo & Kemp 1976). In these experiments, pyridoxal phosphate was reduced onto the enzyme with NaBH$_4$, and several lines of evidence indicated that one mole of pyridoxal phosphate was incorporated almost exclusively in the citrate inhibitory site of the enzyme. The modified enzyme had decreased activity and exquisite sensitivity to ATP inhibition (Fig. 7.3A). Binding studies with labeled ATP showed that the enzyme had enhanced affinity for ATP as would be expected by the synergistic inhibition displayed by ATP and citrate

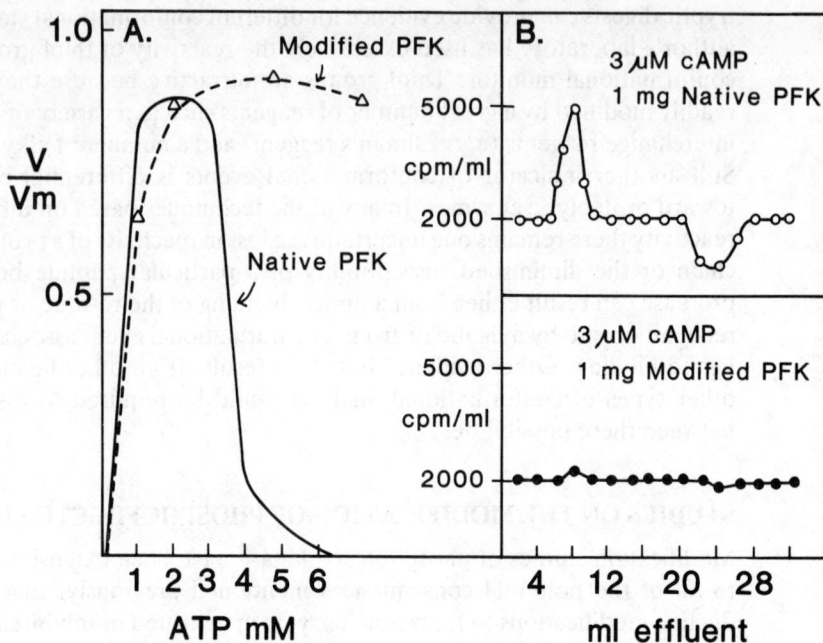

Fig. 7.2 — Effect of F-sulfonylbenzoyladenosine on rabbit muscle PFK. Panel A, ATP inhibition of native (solid line) and modified (dashed lines) PFK. Panel B, Hummel–Dreyer gel filtration binding of cAMP by native and modified PFK. Other conditions as in Gottschalk *et al.* (1983).

(Fig. 7.1C). On the other hand, citrate binding was blocked as shown by the Scatchard plot in Fig. 7.3B. In this instance the enzyme had been modified by the incorporation of about 0.85 moles of pyridoxal phosphate. The data show that what binding remained was undoubtedly due to the unmodified fraction which bound citrate with an affinity equal to that of native enzyme. If the data had shown that the residual binding had identical maximum binding but lower affinity, one would conclude that the modification was not at the citrate site but only influenced the citrate site. This demonstrates the importance of relatively complete binding studies in interpreting such data. The modification by pyridoxal phosphate and sodium borohydride has thus frozen the enzyme in an inhibited conformation by a covalent linkage to a lysine residue in the citrate pocket.

An example of a modification that destroys a binding site can be seen in the experiments of Setlow & Mansour (1970) who modified sheep heart phosphofructokinase with diethylpyrocarbonate. Incorporation of close to four moles of reagent per mole of protomer virtually abolished ATP inhibition. Setlow & Mansour concluded that one or more histidines at the ATP inhibition site were modified. We repeated these experiments with rabbit muscle phosphofructokinase and obtained nearly identical kinetic results (Fig. 7.4A). Fig. 7.4B demonstrates the loss of conformational change to the inhibited conformer following modification as indicated by

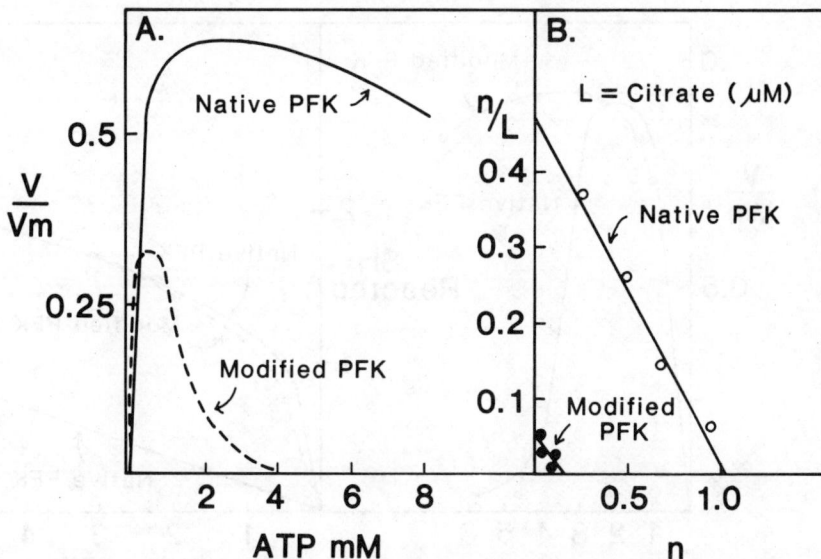

Fig. 7.3 — Effect of modification by pyridoxal phosphate and sodium borohydride. Panel A, effect of modification on ATP inhibition. Data taken from Colombo & Kemp (1976). Other conditions described therein. Note that ATP inhibition of native enzyme is weak because assay was performed at pH 7.35 and high fructose-6-P levels. Panel B, Scatchard plot of [^{14}C]-citrate binding by native and modified PFK. Data of Colombo & Kemp (1976) were replotted.

thiol reactivity. One thiol group of phosphofructokinase is extremely reactive, and this reaction with dithionitro(bis)benzoic acid (DTNB) is inhibited in the presence of ATP. Previous studies have shown that this is a result of ATP binding to an inhibitory site that produces a conformational change (Kemp 1969). It can be seen in Fig. 7.4B that the ethoxyformylated enzyme retains the highly reactive thiol, but its reaction is not inhibited by ATP. These data provide confirmation at the level of allosteric transitions that the interaction of ATP at the inhibitory site has been abolished.

The final potential consequence of chemical modification is typified in a kinetic analysis of phosphofructokinase before and after the modification of the highly reactive thiol group described above. The result is activation of the enzyme at low concentrations of fructose-6-P and sharp reduction in cooperative kinetics (Kemp 1969). This might be interpreted as a blockade of the ATP inhibitory site. The results are not nearly so simple, as indicated by the Hill-type plot shown in Fig. 7.5. This figure shows the two effects of ATP on the kinetics of the unmodified enzyme (open symbols): higher ATP increases the concentration of fructose-6-P necessary to achieve half-saturation and it increases the interaction coefficient. The modified enzyme retains the effect of ATP on half-saturation; thus ATP must continue to interact with the enzyme although the full expression of cooperative behavior has been lost. The interpretation is that binding is not affected but the

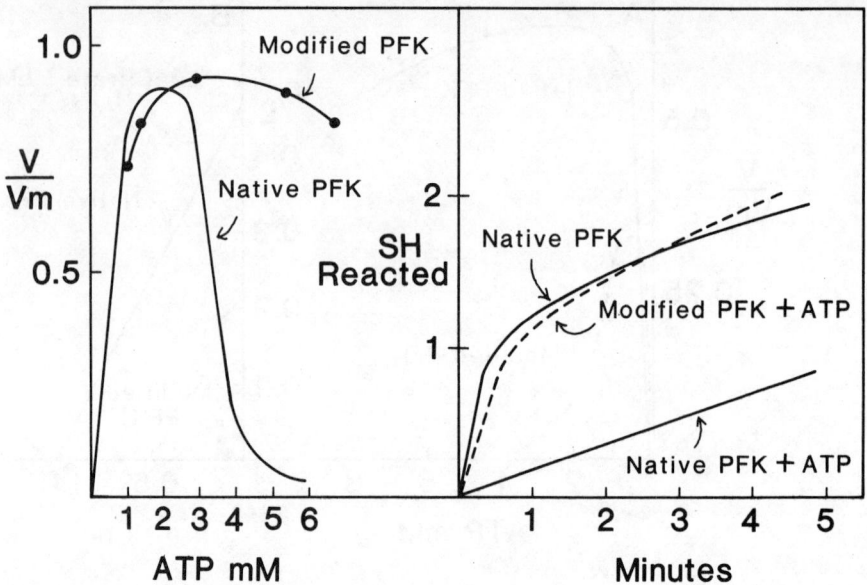

Fig. 7.4 — Effect of ethoxyformylation of PFK. Modification of rabbit muscle PFK performed as described by Setlow & Mansour (1970). Panel A, ATP inhibition of native and modified PFK. Panel B, thiol reactivity of native and modified PFK. Conditions and concentrations for the reaction of protein sulfhydryl groups with DTNB are identical to those described by Foe *et al.* (1983).

conformational change is partially blocked. This is consistent with the previously mentioned data that this thiol group is an indicator of the active–inactive transition and must occupy a site that undergoes change during conformational events. One can imagine that a bulky reagent attached to this thiol may inhibit the conformational change. To complete this study and verify such conclusions, ligand binding of the native and thiol-modified enzyme should be performed.

LIMITED PROTEOLYSIS AS A TOOL FOR SPECIFIC MODIFICATION

Because allosteric enzymes are usually large, multi-domain enzymes with regulatory sites located on domains often separated from the catalytic site, limited proteolytic digestion is a valuable tool for the modification of allosteric properties, particularly in those instances where substantial information is available on the primary structure of the enzyme. Limited protelysis is frequently used for the domain analysis of complex proteins. An excellent example of this is the work from Wakil's laboratory on the fatty acid synthetase (Mattick *et al.* 1983). In that instance, digestion studies with a battery of proteases allowed mapping of domains and led to studies that

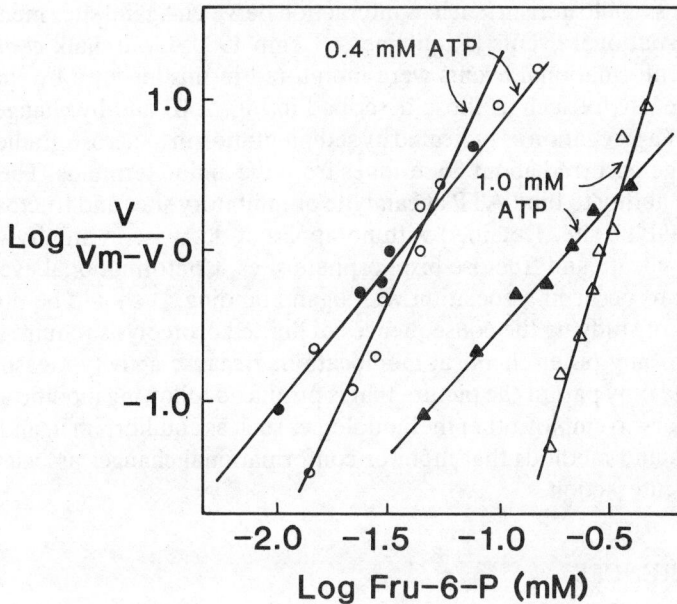

Fig. 7.5 — Effect of thiol modification on PFK. Hill-type plot of native (open symbols) and modified (closed symbols) PFK at two different concentrations of ATP. Thiol modification was carried out with Ellman's reagent (DTNB). Data of Kemp (1969) have been replotted.

involved direct isolation of active fragments for several of the seven enzymatic activities.

The use of proteolysis to study the properties of allosteric enzymes actually has a very long history. An early study of glycogen phosphorylase showed a proteolytic modification of allosteric properties by trypsin. Many other allosteric enzymes have been modified by proteolysis. Phosphofructokinase is included in his list. Clues to the existence of two 'super domains' of phosphofructokinase was shown by the work of Emerk & Frieden (1974) who cleaved phosphofructokinase into approximate halves. Surprisingly, this modification produced no measurable change in catalytic and regulatory properties. Later it was shown (Kemp et al. 1981) that, in addition to cleavage in the middle of the 83k protomer, the trypsin treatment removed 8 residues of the amino terminus and 6 residues from the carboxyl terminus, again without changing the properties of the enzyme. On the other hand, limited exposure of phosphofructokinase to S. aureus V8 protease removes 17 residues from the carboxyl terminus and dramatically decreases inhibition by ATP (Valaitis et al. 1986). Three histidine residues are found in this fragment, and it may include one or more of those histidines implicated by Mansour's laboratory in the binding of ATP to the inhibitory site (Setlow & Mansour 1970).

Under appropriate conditions, modification of phosphofructokinase by subtilisin led to inactivation of the enzyme while the protein retained features of allosterism, such as interaction between ligand sites mediated by conformational events (Riquelme & Kemp 1980, Gottschalk *et al.* 1983). The conformational events were monitored in this instance by changes in thiol reactivity such as those described in Fig. 7.4B and by changes in the state of aggregation as indicated by sedimentation in a sucrose gradient. This cleavage occurred about 60 residues from the amino terminus. The protein lost its ability to bind ATP at catalytic or inhibitory sites and fructose-6-P at the catalytic site. Retained with no apparent decrease in affinity were the AMP, citrate and fructose bisphosphate sites. Conformational events continued to occur in association with ligand binding. It should be noted that means of studying the consequences of limited proteolysis require identical tools to any other chemical modification. Because activity measurements provide only part of the picture that is produced following modification, it is necessary to employ other methodologies such as equilibrium ligand binding studies and methods that monitor conformational changes associated with ligand interaction.

REFERENCES

Changeux, J.P., Gerhart, J.C. & Schachman, H.K. (1968) *Biochemistry* **7** 1774–1780.

Colombo, G. & Kemp, R.G. (1976) *Biochemistry* **15** 530–532.

Colowick, S.P. & Womack, F.C. (1969) *J. Biol. Chem.* **244** 774–777.

Emerk, K. & Friedman, C. (1974) *Arch. Biochem. Biophys.* **164** 233–240.

Foe, L.G., Latshaw, S.P. & Kemp, R.G. (1983) *Biochemistry* **22** 4601–4606.

Gottschalk, M.E., Latshaw, S.P. & Kemp, R.G. (1983) *Biochemistry* **22** 1082–1087.

Hayes, J.E., Jr. & Velick, S.F. (1954) *J. Biol. Chem.* **207** 7352–7357.

Hummel, J.P. & Dreyer, W.J. (1962) *Biochim. Biophys. Acta* **63** 530–532.

Kemp, R.G. (1969) *Biochemistry* **8** 4490–4496.

Kemp, R.G., Foe, L.G., Latshaw, S.P., Poorman, R.A. & Heinrickson, R.L. (1981) *J. Biol. Chem.* **256** 7282–7286.

Kemp, R.G. & Foe, L.G. (1983) *Mol. Cell. Biochem.* **57** 147–154.

Kitajima, S. & Uyeda, K. (1983) *J. Biol. Chem.* **258** 7352–7357.

Liou, R.-S. & Anderson, S.R. (1978) *Biochemistry* **17** 999–1004.

Mansour, T. & Colman, R. (1978) *Biochem. Biophys. Res. Commun.* **81** 1370–1376.

Mattick, J.S., Tsukamoto, Y., Nickless, J. & Wakil, S.J. (1983) *J. Biol. Chem.* **258** 15291–15299.

Paulus, H. (1969) *Anal. Biochem.* **32** 91–100.

Riquelme, P.T. & Kemp, R.G. (1980) *J. Biol. Chem.* **255** 4367–4371.

Setlow, B. & Mansour, T. (1970) *J. Biol. Chem.* **245** 5524–5531.

Uyeda, K. (1979) *Adv. Enzymol Relat. Areas Mol. Biol.* **48** 193–244.

Valaitis, A., Foe, L.G. & Kemp, R.G. (1986) *Fed. Proc.* **45** 1808.

8

Active-site studies by secondary structure prediction

Dr Hilda Cid, Laboratorio de Biofísica Molecular, Universidad de Concepción, Concepción, Chile

STATEMENT OF THE PROBLEM

In order to understand the mechanism of action of an enzyme, a hormone or a macromolecular carrier, it is necessary to determine, as precisely as possible, the tertiary structure of the protein and, when relevant, its quaternary structure. Since the first tertiary structure of a protein was determined about 30 years ago, many important developments in the biological sciences have been a direct consequence of the advances in the knowledge of the structure of biological macromolecules.

However, the determination of the tertiary structure of a protein (performed by X-ray diffraction) has not yet become a routine task. Several problems, such as growing protein crystals, or searching for heavy atom derivatives, sometimes take several years before they are adequately solved.

The secondary structure of a protein describes the spatial folding of the polypeptide chain which, in this specific way, acquires a more stable conformation. The process of folding allows the shielding of the hydrophobic amino acid residues in the interior of the molecule, whereas the hydrophilic ones are exposed to the surrounding solvent. The folding of the polypeptide chain also facilitates the proximity of residues capable of forming stabilizing bonds such as disulfide bridges or hydrogen bonds between the turns of a helical structure or between the strands of a β-pleated sheet.

For several years, the secondary structure of a protein was obtained as a by-product of tertiary structure determination. It provides a simpler way to describe the complicated architecture of the macromolecule than the spatial location of the thousands of atoms, or the hundreds of amino acid residues. Figs 8.1a and b illustrate the presentation of the structure of carbonic anhydrase B by a secondary structure model, and by a chain containing just

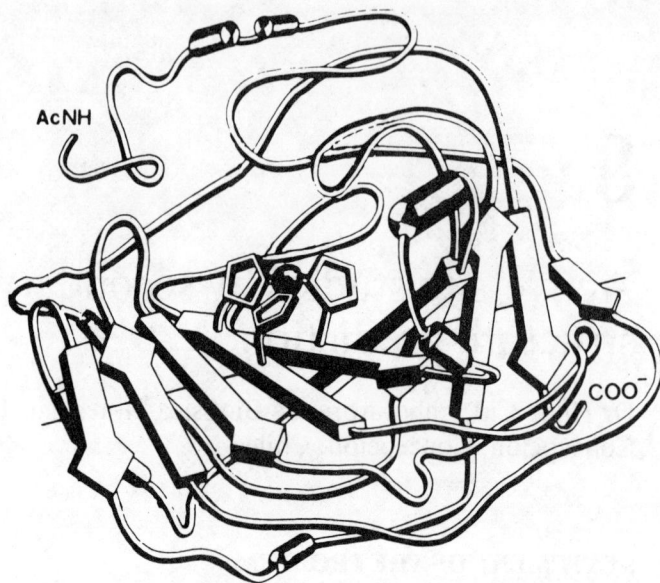

(a)

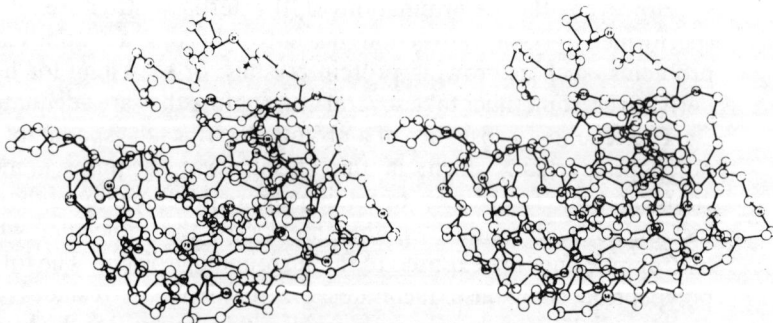

(b)

Fig. 8.1 — (a) Secondary structure of human carbonic anhydrase B (HCAB), in the cylinder and arrow representation. (b) Tertiary structure of HCAB. Stereoscopic computer drawing of the carbons. (From Nostrand 1974.)

the α carbon of each amino acid residue. It is only in the last 15 years that the determination of the secondary structure, without the previous knowledge of the tertiary structure, has been considered possible.

The discovery of the genetic code and the determination of DNA sequences showed that the primary structure of a protein is the only information genetically transcribed. This fact means that the amino acid sequence already contains all the necessary information to determine the folding of the polypeptide chain; that is, it determines both the secondary and the tertiary structure. This conclusion is supported by denaturation–renaturation experiments on enzymes.

Following this principle, several methods have been designed to predict the secondary structure of a protein from a knowledge of its amino acid sequence. Some of these methods are based on empirical probabilities (Chou & Fasman 1974), numerical algorithms (Lim 1974), or a combination of physicochemical measurements from a data base of known protein structures (Cid *et al*. 1982).

The possibility that the secondary structure of a protein could be predicted in the absence of a knowledge of the tertiary structure opens new trends in enzyme active-site studies. Even if the complete enzyme structure cannot be known precisely, the location of the active-site and some information about its morphology, would be a valuable tool in the comprehension of its mechanism of action.

It will be shown in three examples given below, that in the case of enzyme families with known primary structures, it is possible to obtain important information about the active-site by a combination of secondary structure prediction and chemical studies. It is even possible to propose a spatial folding of the polypeptide chain by a combination of secondary structure prediction and model building.

The central idea is that the active-site region in a family of enzymes should present a much higher degree of structural invariability in the secondary and tertiary structure than is usually found in the primary structures. The shape of the active 'pocket' must be conserved since all enzymes accommodate the same substrate or part of the substrate within it. This type of approach has been used in the study of three protein families: snake venom phospholipases A_2, β-lactamases obtained from different micro-organisms, and fructose-1,6-bisphosphatase from various animal tissues.

PREDICTION OF THE SECONDARY STRUCTURE OF A PROTEIN

Of the several methods designed to predict the secondary structure of a protein from the amino acid sequence, none has proved to be 100% reliable when applied to proteins whose tertiary structures (and therefore, secondary structures) have been fully determined by X-ray diffraction methods. Success is variable. Two methods, however, one proposed by Chou & Fasman (1974) and the other by Cid and collaborators (1982), have been reported to give 80% reliability when applied to globular proteins.

Chou and Fasman's method for secondary structure prediction

The Chou and Fasman method is based on empirical probabilities. It defines conformational parameters P_α, P_β and P_t for each of the 20 natural amino acids. These parameters represent the probability which each amino acid has of participating in a helix, a β-structure or a turn structure. They are, in fact, the normalized frequency of occurrence of each amino acid residue in that particular type of structure, as established from a data base of 29 fully determined protein structures. A probability average greater than 1.0, obtained for a group of amino acids taken in sequence (6 for a helix, 5 for a β-strand, and 4 for a β-turn) is an indication that a specific type of structure is likely to occur in that region of the sequence. To improve the sensitivity of the method in the vicinity of the limit value 1.0, the probability average can be replaced by a product of the conformational parameters (Dufton & Hider 1977). This and two other modifications of the method, one that considers four conformational parameters for each amino acid residue in a turn structure (Chou & Fasman 1978) and another that allows a differentiation between β-strands participating in parallel or antiparallel β-structures (Lifson & Sander, 1979), have proved to be very useful in the secondary structure prediction.

Hydrophobicity profiles' method for secondary structure prediction

The hydrophobicity profiles method combines physico-chemical measurements of the solubilities of amino acids in polar and non-polar solvents with information obtained from a data base of 21 known protein structures (Cid et al. 1982). The method gives the relative position of portions of the polypeptide chain with respect to the protein surface since a linear correlation exists between the 'surrounding hydrophobicity' H_f, as defined by Ponnuswamy et al. (1980), and the average distance of an amino acid residue to the protein surface, as measured on 21 known protein structures. The 'surrounding hydrophobicity' for a given residue is defined as the sum of the Tanford hydrophobicities of all the amino acids included in an 8 Å-radius sphere, centered at the α-carbon of the residue in consideration. Since the calculation of the surrounding hydrophobicity implies the knowledge of the tertiary structure, a 'bulk hydrophobic character' $\langle H_f \rangle$, has been calculated as an experimental average value from the tertiary structures available for 21 proteins.

The hydrophobicity profile is simply a plot of $\langle H_f \rangle$ versus the amino acid number in the sequence. Four basic profiles have been defined for 4 types of secondary structure: helix, β-turn, and buried, and exposed β-strands. These basic profiles are represented in Fig. 8.2. The identification of these basic patterns in the hydrophobicity profile of the protein yields the predicted structure.

Model building of predicted secondary structures

In order to obtain as much information as possible from a predicted secondary structure, three-dimensional models can be built using rigid arrows, cylinders and 'hair pins' to represent β-strands, α-helices and

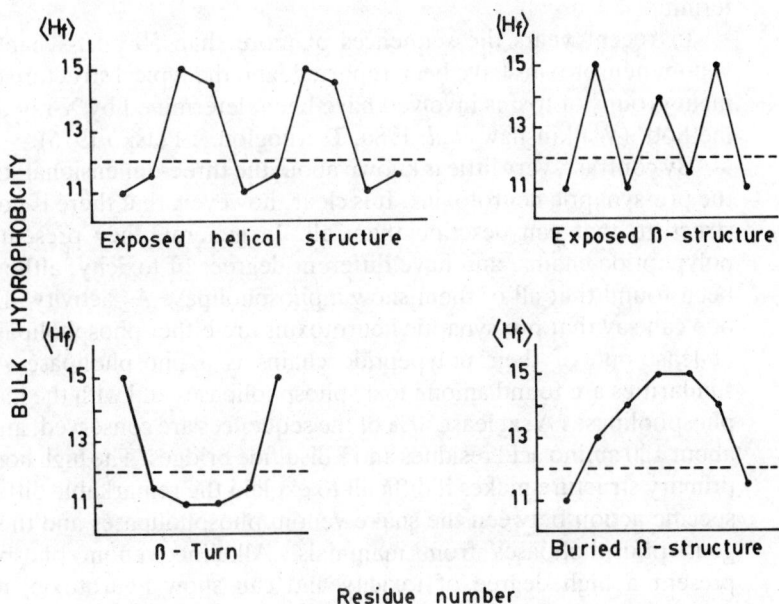

Fig. 8.2 — The four basic hydrophobicity profiles. The number of the amino acids in the sequence has been plotted against the 'bulk hydrophobic character'.

β-turns, respectively. These elements are joined by mobile connections and by flexible wire that represent random coil regions. The lengths of the building elements are scaled to the number of amino acids involved, and to the distance between α-carbons in that particular type of structure. In building models, the following complementary information must be considered:

(1) Distinction between exposed and buried β-strands, as predicted by the hydrophobicity profiles.
(2) Preferences for β-strands to be part of a parallel or antiparallel β-sheet, as given by the modified Chou & Fasman method.
(3) Stabilization of helical and β-structures in one of the three super secondary structures defined by Levitt & Chothia (1976): αα, ββ, βαβ.
(4) Proximity of some amino acid residues based on chemical evidence. Examples of these models are given in Figs 8.6 and 8.8.

Search for a 'Toxic Site' in Snake Venom Phospholipases A₂

Protein toxins from snake venom block transmission across the cholinergic neuromuscular junction, either by a post-synaptic, curare-like action on the nicotine acetylcholine receptors of the muscle end plate, or by a presynaptic interference with the release of acetylcholine from the motor nerve terminals.

In recent years the sequences of more than 50 post-synaptic snake-venom neurotoxins have been reported, and the typical structure of the two main groups of toxins involved have been determined by X-ray diffraction method. (Walkinshaw *et al*. 1980, Tsernoglou & Petsko 1976).

By contrast, very little is known about the three-dimensional structure of the pre-synaptic neurotoxins. It is clear, however, that there is no common structure that can describe them all. In general, they present multiple polypeptide chains, and have different degrees of toxicity, although it has been found that all of them show a phospholipase A_2 activity. Therefore, one can say that pre-synaptic neurotoxins are either phospholipases A_2, or at least one of their polypeptide chains is a phospholipase A_2. Great similarities are found among toxic phospholipases and with the mammalian phospholipases A; at least 50% of the sequences are conserved, and all have about 120 amino acid residues and 7 disulfide bridges. The high homology in primary structure makes it difficult to explain the remarkable differences in specific action between the snake venom phospholipases and the homolo-gous phospholipases from mammals. Alkaline venom phospholipases present a high degree of toxicity and can show neurotoxic, mitotoxic, cardiotoxic, or anticoagulant activity; acidic or neutral phospholipases A are less toxic or non-toxic. It is reasonable, then, to look for differences in the secondary structures that could account for their toxic action.

The relationship between secondary structure and toxicity is the aim of a preliminary study on the secondary structure of the following toxins: notexin, obtained from the venom of *Notechis scutatus scutatus*, which consists of a single polypeptide chain of 119 amino acid residues with a highly lethal neurotoxic action plus miotoxic activity; phospholipase A_2 from the venom of *Haemachatus haemachatus* which only has phospholipase activity; and chain A from *Bungarus multicintus* neurotoxin with a high degree of lethality (Arriagada & Cid 1980). The secondary structures, predicted both by the Chou & Fasman method and by the hydrophobicity profiles method, were compared to the predicted and observed secondary structure of bovine phospholipase A_2, whose tertiary structure has been fully determined by high resolution X-ray diffraction (Dijkstra *et al*. 1981). The predicted secondary structure of the mammalian phospholipase, compared to that obtained from the tertiary structure, gave an estimation of the reliability of the method in this case. Fig. 8.3 illustrates the relationships between predicted secondary structures of the three toxins among themselves and with the predicted and the X-ray determined secondary structures of bovine phospholipase A_2. The helical zones have been represented by springs and the zones of extended structure (that form part of β-structures) by stippled blocks. A remarkable similarity of the four enzymes can be seen in the first part of the sequence up to residue 60. Important differences are observed in the second half of the sequence: the three toxins present mainly extended structures not found in the predicted or in the experimentally determined structures of the mammalian enzyme.

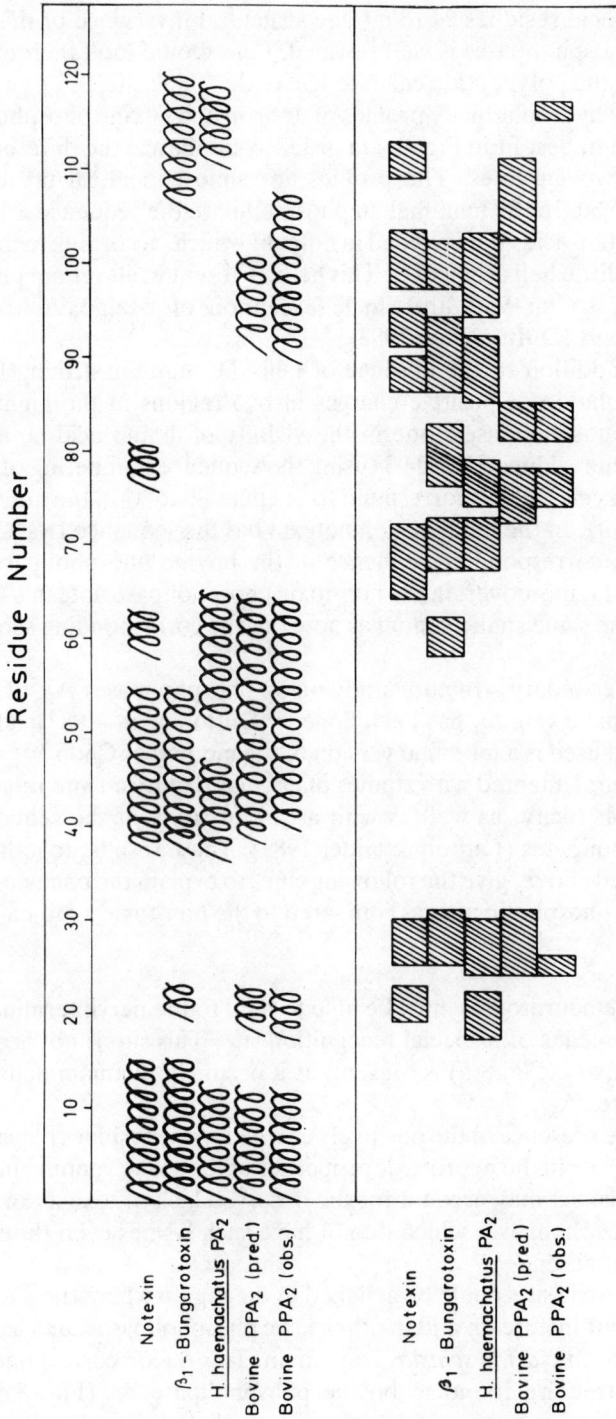

Fig. 8.3 — Structural relationships between phospholipases. The predicted secondary structures of three snake venom phospholipases A_2 and of the bovine enzyme have been compared with the secondary structure of the latter obtained from its known tertiary structure. Helical zones are represented by springs, and regions of β-structure by stippled blocks.

Since the catalytic site of phospholipases A_2 has been located between amino acid residues 24 to 60, the structural invariance of this zone on the four phospholipases is well justified. One should look therefore in the last part of the polypeptide chain to locate the 'toxic site'.

The hydrophobicity profiles of notexin and bovine phospholipase A_2 are aligned to best fit in Fig. 8.4 in order to emphasize the differences between these two enzymes. The profiles are almost identical up to amino acid residue 60. In the mammalian phospholipase the sequence is then followed by 5 amino acids not included in notexin, which, according to the X-ray data, form a little helix (helix D). This helix is absent in all venom phospholipases studied so far with the single exception of τ-taipoxin from *Oxyranus scutellatus* (Dufton *et al.* 1983).

In addition to the absence of helix D, another striking feature is the accumulation of positive charges in two regions of the highly neurotoxic basic phospholipases: one in the vicinity of amino acid 60 and the other between residues 85 and 93 (using the sequence numbering of the mammalian enzyme) which correspond to residues 80 to 87 in the notexin sequence (Fig. 8.4). In the latter region notexin has the sequence Lys 82, Lys 83, Lys 84; the corresponding sequence in the bovine phospholipase being Asn, Asn, Ala; moreover, in the non-toxic phospholipase notechis II-1, obtained from the same snake venom as notexin, the corresponding sequence is Lys, Tyr, Gly.

A secondary structure study on 32 phospholipases A_2, of which 29 are from snake venom, has been done by Dufton *et al.* (1983). The prediction method used is a modified version of the method of Chou & Fasman, which was complemented with studies of circular dichroism and relative interface hydrophobicity, as well as with a careful study of the sequences of such phospholipases (Dufton & Hider 1983). These results, together with those reported above, give the following clues to explain the behavior of the snake venom phospholipases, as compared to the non-toxic mammalian phospholipases A_2:

(1) The neurotoxins must be able to bind to the nerve terminal membrane by means of a special recognition site. This site is not accessible when helix D (Fig. 8.5) is present, as it occurs in the mammalian phospholipases A_2.

(2) The presence of the positively charged lysine residues plays a fundamental role in the neurotoxic properties of the snake venom phospholipases. This fact may account for the observed lack of toxicity of some venom phospholipases which do not have such lysines even though helix D is also absent.

(3) A 'toxic site' could be assigned to the region where the 3 lysine residues occur in notexin and in other toxic phospholipases such as notechis II-5 and *Enhydrina schistosa* myotoxin. This region corresponds to the only β-structure found in bovine phospholipase A_2 (Fig. 8.5) and to an extensive zone of β-strands predicted for the snake venom phospholipases.

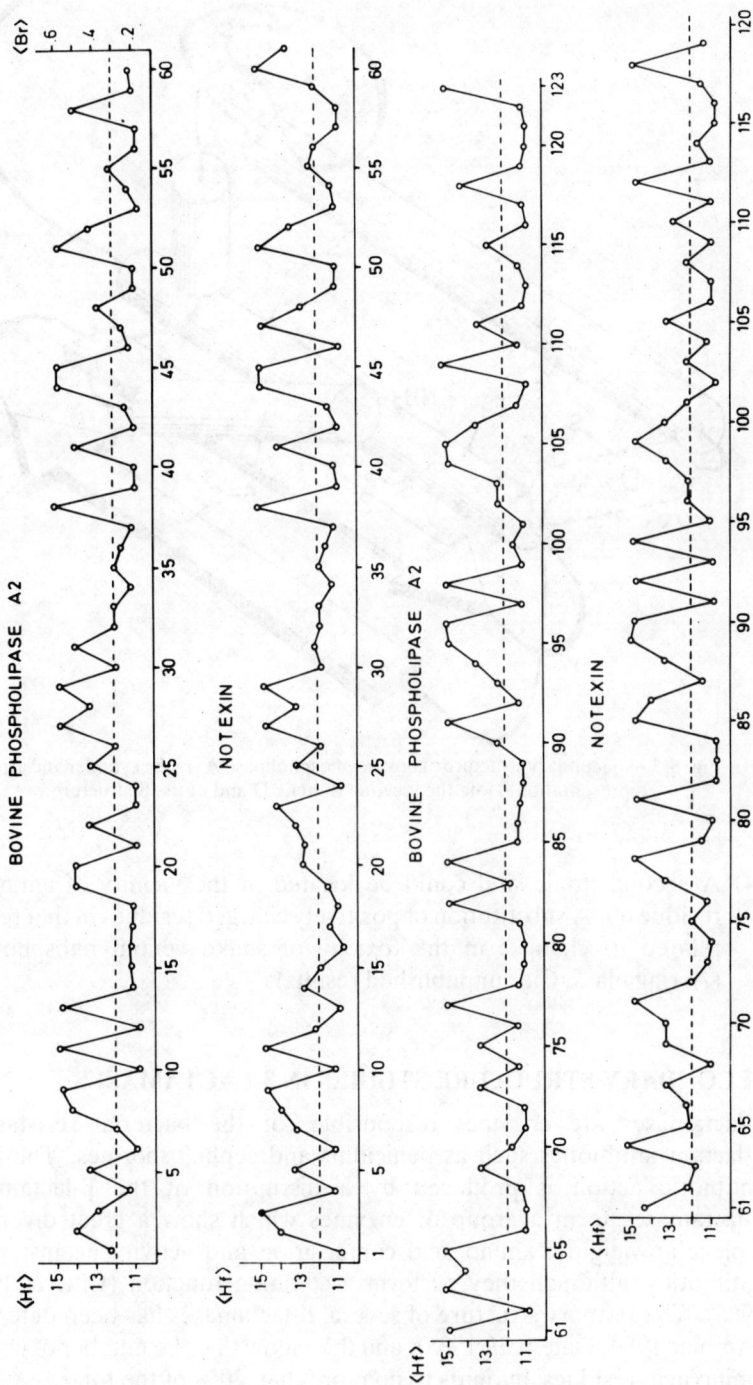

Fig. 8.4 — Hydrophobicity profiles of notexin and bovine phospholipase A₂ aligned for best fit. Note the similarity in predicted secondary structures from amino acid 1 to 60.

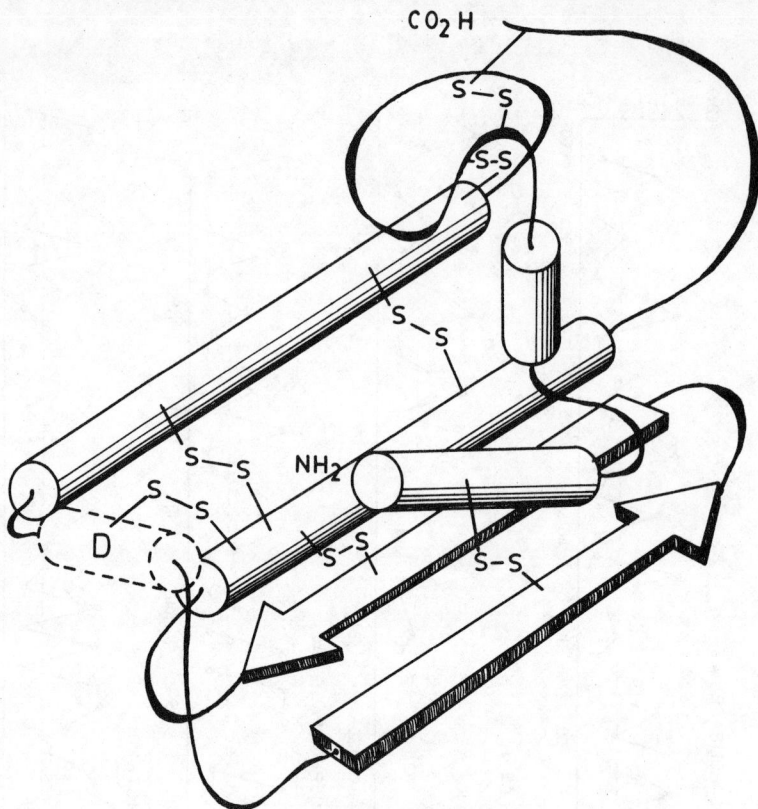

Fig. 8.5 — Secondary structure of bovine phospholipase A_2 in the cylinder and arrow representation. Note the location of helix D and of the β-structure.

(4) A second 'toxic site' could be located in the vicinity of amino acid residue 60. A substitution of positively charged residues in that region is related to changes in the toxicity of snake venom phospholipases (Arriagada & Cid, unpublished results).

SECONDARY STRUCTURE STUDIES IN β-LACTAMASES

β-lactamases are enzymes responsible for the bacterial resistance to β-lactam antibiotics such as penicillins and cephalosporines. The loss of antibiotic action is produced by a disruption of the β-lactam ring. β-lactamases form a group of enzymes which show a great diversity in molecular weights, amino acid composition and activity against various antibiotics, although they perform a common function (Citri & Pollock 1966). The primary structure of several β-lactamases has been determined (Ambler 1980, Dale *et al.* 1985), and they show that the number of invariant amino acid residues amounts to no more than 20% of the total.

It is obvious that a precise knowledge of the mechanism of action of these enzymes would be of the greatest importance for medical purposes. Therefore, much effort has been spent trying to solve the three-dimensional structure of β-lactamases and of their active-sites. Although some low resolution X-ray diffraction data have been reported, the tertiary structure of none of these enzymes has yet been completely solved.

In the absence of tertiary structure information, knowledge of secondary structure can provide information about the conformation of the active-site, and, at the same time, it can furnish a better pattern for comparison than does the primary structure. One would expect a structural homology far better than 20% in the vicinity of the active-sites of enzymes that act upon the same substrate.

A common model for the structure of four β-lactamases

Bunster & Cid (1984) made a secondary structure prediction for the β-lactamases obtained from *Escherichia coli*, *Bacillus cereus*, *Bacillus licheniformis*, and *Staphylococcus aureus*, using both prediction methods previously described. The primary structures were those reported by Ambler (1980). The first important result that came out of the predictions was that on the average, 56% of all amino acid residues were involved in zones with a conserved secondary structure. Models of the four secondary structures predicted were built, according to the technique described before. These models are represented in Fig. 8.6.

The models point to various interesting results:

First, the four β-lactamases can be described as two-domain structures, consistent with the report for the *E. coli* enzyme from a low resolution structure obtained by X-ray methods (Knox *et al.* 1976).

Second, a large part of the secondary structure is conserved, at least in three of the four enzymes. This corresponds to the shadowed structural elements in Fig. 8.6. The constant structures may differ in the lengths of the β-strands, helices or random coiled regions, but they are still easily recognized. The small number of invariant amino acid residues present in a zone with a conserved secondary structure is remarkable, as pointed out in Table 8.1.

Third, the region containing residues 31 to 157 presents 84% of all its amino acids involved in invariant secondary structures, while only 22% of the sequence is conserved. This structural constancy, and the fact that this domain includes all the amino acid residues that seem to have something to do with the activity of the enzymes, could be an indication that the active-site, and presumably all the catalytic function of these β-lactamases, is restricted to the first domain. Even if domain I controls the catalytic action, it could still be possible that domain II could cooperate with the catalytic site since the flexibility of the inter-domain region would allow a close contact. The existence of β-lactamases with a molecular weight similar to that of domain I, as is the case for the enzymes obtained from *Pseudomonas aeruginosa* plasmids R 157 and Rms 149 (Matthew 1978, Ambler 1980), and

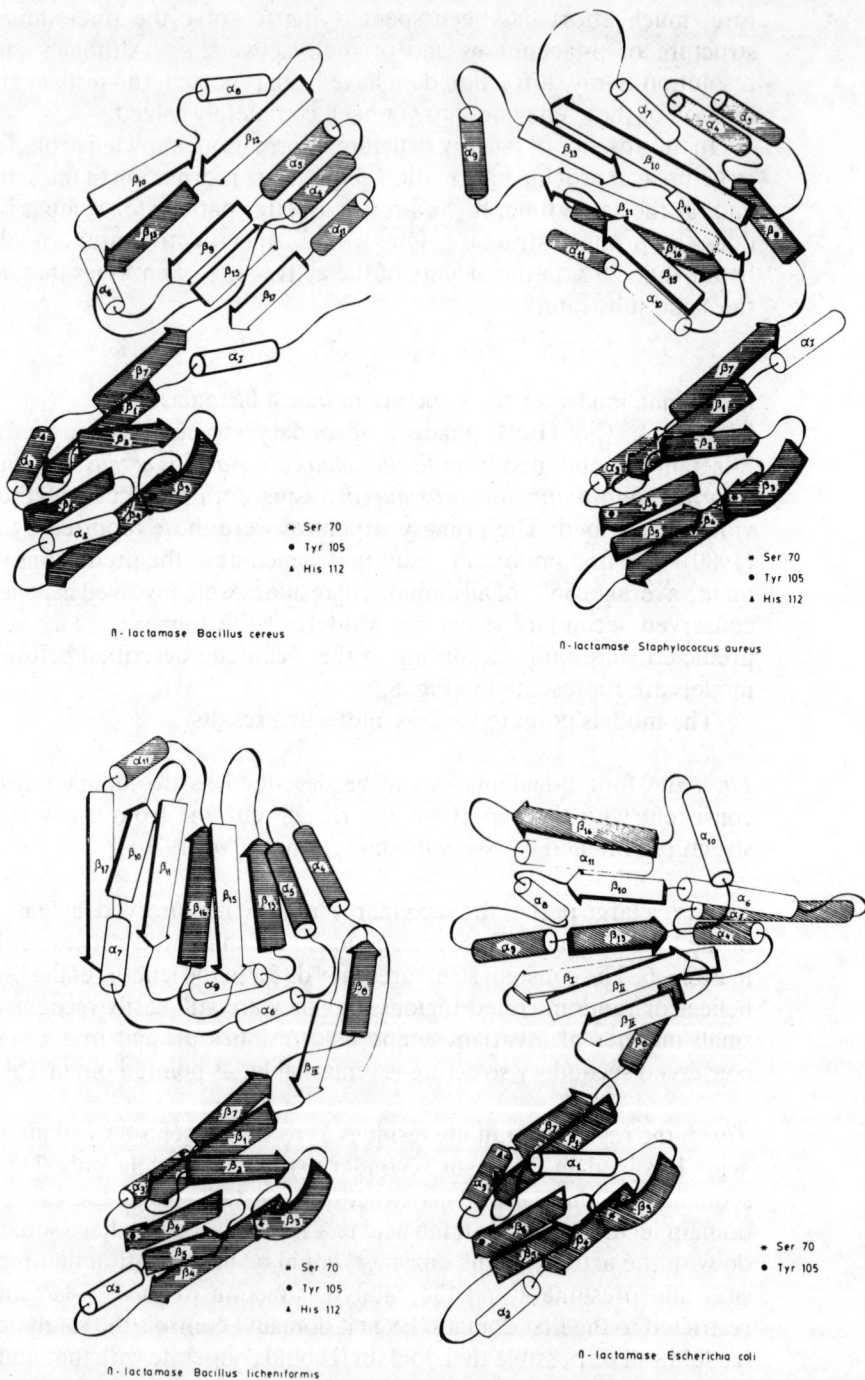

Fig. 8.6 — Models for the predicted secondary structures of four β-lactamases. The shadowed structural elements are those invariant in at least three of the four structures. (From Bunster & Cid 1984.)

Table 8.1

Regions with a conserved secondary structure in at least three of the four β-lactamases studied.

Region	AA residues	Invariant residues	Structure involved
1	39 to 81 (43)	9	$r-\beta_1-\beta_2-\beta_3-r-\beta_4$
2	94 to 157 (64)	18	$\beta_5-t-\beta_6-\alpha_2-r-\beta_7-r-\beta_8-r-\alpha_3-r-\alpha_4$
3	176 to 183 (8)	2	r
4	227 to 282 (546)	7	$\beta_{13}-r-\beta_{15}-r-\beta_{16}-r-\alpha_{11}$

β = β-strand   α = α-helix   t = β-turn   r = random coil
The proposed secondary structure is depicted in Fig. 8.6. The total number of amino acid residues included in the zones of invariant secondary structure is indicated in brackets.

from Streptomyces UCSM-104 (Campos *et al.* 1979) supports a model with the β-lactamase activity restricted to domain I.

Fourth, in the model proposed, the active-site would be formed by parts of the following constant regions: β_1, β_3, β_4, and a charged loop previous to β_3. This loop contains Ser 70 which participates in the catalytic reaction (Waxman Strominger 1980). Both Tyr 105 and His 112, which should be near the active-site (Durkin & Viswanatha 1980) are also located in this neighborhood. The presence of charges near Ser 70 coincides with the reported existence of carboxylic groups in the active-site (Durkin & Viswanatha 1980).

Test of the validity of the model
To test the validity of the model, two aspects have been considered: (a) if the activity of the enzyme is restricted to domain I, it should be maintained in the absence of domain II; and (b), a three-dimensional model of domain I should reveal a site compatible with the substrate and with the catalytic reaction.

(a) The generation of an active domain I
To prove the first point, *B. cereus* β-lactamase (molecular weight 27 800) was reacted with CNBr under very mild conditions. CNBr reacts primarily with methionines to cause cleavage, and reacts secondarily with cysteines. The molecule chosen has no cysteines and only four methionines, none of them in domain I, so this reaction seemed adequate to separate domain I. The reaction was performed in phosphate buffer pH 6.8, with 50% molar excess 6M CNBr in HCl, for 4 hours. These conditions were chosen in order to protect, as much as possible, the enzymatic activity. Polyacrylamide gel electrophoresis of the enzyme after CNBr treatment showed two active bands, corresponding to peptides of molecular weights 28 000 and 14 000 (Martínez 1984, Martínez & Cid 1984). This result is consistent with a partial

splitting of the enzyme at Met 130, which would give rise to two peptides of very similar molecular weight of about 14 000. These peptides were not resolved in the SDS polyacrylamide electrophoresis. The larger active peptide corresponds to the native enzyme. This interpretation of the results was confirmed when the peptide of molecular weight 14 000 was neatly resolved into two bands by polyacrylamide-urea gel electrophoresis.

These experiments have shown that the *B. cereus* β-lactamase can be split at Met 130 giving rise to two peptides. At least one of them remains active, and one of them contains the intact domain I of the proposed structure for β-lactamases. Since only domain I contains Ser 70 which participates in the catalytic mechanism, the active peptide must be domain I.

(b) Location of a possible active-site
A three-dimensional model of domain I of the β-lactamase from *B. cereus* was built using Kendrew skeletal models, following the pattern of the proposed structure represented in Fig. 8.6. In this model a cavity, presumably the active-site, is clearly visible. This site can accommodate a molecule of benzyl penicillin as substrate. The substrate can be stabilized in a unique orientation. The benzene ring interacts hydrophobically with the aromatic rings of two phenylalanines, whereas the carboxyl group of the five-member ring can make a hydrogen bond with a lysine (Fig. 8.7). The substrate oriented in this way is sufficiently close to Ser 70 to allow molecular interaction with it (Bunster 1984). This site, found in the *B. cereus* enzyme, should be present at least in the four β-lactamases studied and it should be the same for all β-lactamases type I since they catalyze the same reaction. The changes in affinities for different β-lactam substrates that have been reported can be explained by the sequence differences found in the region that contributes to stabilize the binding of the substrate.

ACTIVE-SITE STUDIES OF FRUCTOSE-1,6-BISPHOSPHATASE FROM PIG KIDNEY

Fructose-1,6-bisphosphatase catalyzes the hydrolysis of fructose-1,6-bis-phosphate to fructose-6-phosphate and inorganic phosphate, a reaction which is a key regulatory step in gluconeogenesis. The enzyme is inhibited allosterically by AMP and also by fructose-2,6-bisphosphate..

Even though the enzyme, isolated from different gluconeogenic tissues, has been extensively studied (Pontremoli & Horecker 1970), structural data regarding the active-site and regulatory sites are far from complete. Some amino acid residues involved in the active-site and in the AMP binding site have been identified in the enzyme obtained from rabbit liver and, by sequence homology, the identification has been extended to the pig kidney enzyme (Marcus *et al.* 1982).

The complete primary structures have been determined for the enzymes from sheep liver (Fischer & Thompson 1983) and pig kidney (Marcus *et al.* 1982). Partial sequences are known for enzymes obtained from rabbit, chicken, turkey and mouse liver, and rabbit kidney (Pontremoli *et al.* 1983).

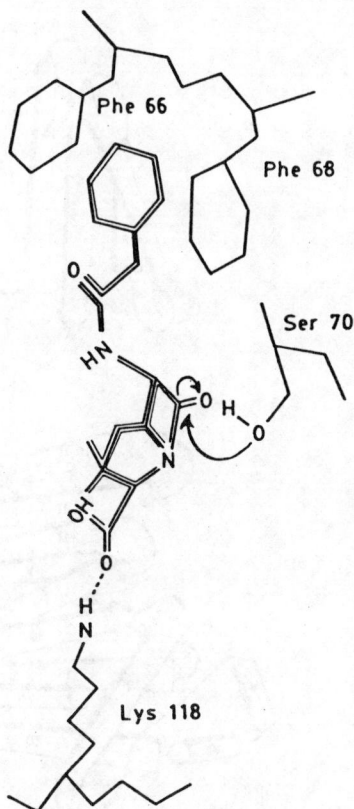

Fig. 8.7 — Stabilization of the substrate benzylpenicillin in the proposed active-site for *B. cereus* β-lactamase.

No tertiary structure has been reported to date for any of these enzymes, even though crystals have been obtained for the chicken (Anderson & Matthews 1977) and rabbit liver (Soloway & McPherson 1978), and for the pig kidney enzymes (Seaton *et al.* 1984).

A model for the structure of fructose-1,6-bisphosphatase from pig kidney
This enzyme is a tetramer formed by subunits of molecular weight 36 534, as calculated from the 335 amino acid sequence. Secondary structure studies were made by the methods already discussed; a secondary structure was proposed and a model built based on this structure (Martinez & Cid 1986) shown in Fig. 8.8.

The predicted secondary structure has a 38% β-structure, 22% helical structure and 4% β-turns. The enzyme subunit is organized in two domains, as would be expected for a globular protein of structural class α/β with more than 300 amino acids (Levitt & Chothia 1976). Each domain can be described as a β-barrel formed by parallel and antiparallel β-strands.

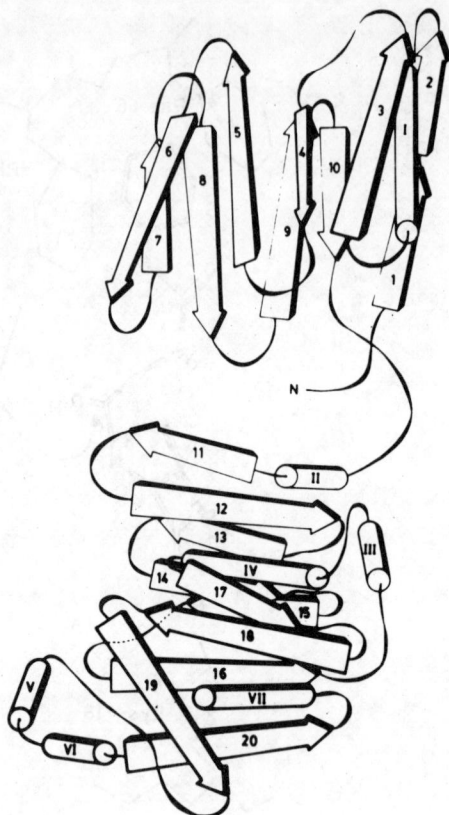

Fig. 8.8 — Two-domain model for the structure of fructose-1,6-bisphosphatase from pig kidney, in the cylinder and arrow representation. (From Martínez & Cid 1986.)

Domain I is formed by the first 141 residues, followed by a 17-residue strand of random coiled structure and by domain II including amino acids 157 to 335.

Location of the active and regulatory sites

The location of the active-site, and of the AMP site in this model, was established by secondary structure homology with the enzyme obtained from rabbit liver, where the sequences in the neighborhood of these sites are known (Xu *et al*. 1982, Suda *et al*. 1982). The secondary structure comparison is easily made by aligning the hydrophobicity profiles. In Fig. 8.9, the region from amino acid residues − 79 to − 39 of the rabbit enzyme which has been postulated to involve the active-site is aligned with the sequence 249 to 290 of the pig kidney enzyme. The negative numbers of the rabbit enzyme sequence, proposed by Xu *et al*. (1982), describe the position of each amino acid residue in a fragment of $Mr = 9,850$ with respect to the COOH terminus of the enzyme included in the peptide. There is a 75%

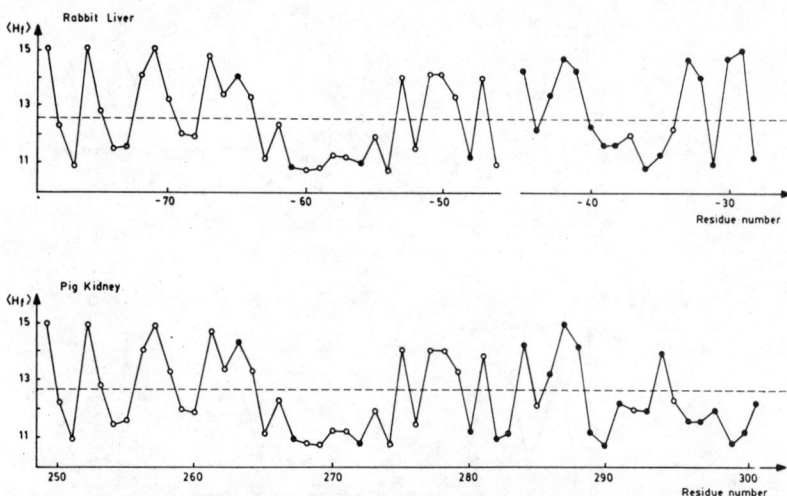

Fig. 8.9 — Hydrophobicity profiles for the sequences -79 to -27 of rabbit liver and 249 to 301 of pig kidney fructose-1,6-bisphosphatases aligned for best fit. $\langle H_f \rangle$ values for non-conserved amino acid residues are represented with filled circles. Lys -54 and its analog Lys 274, which should be part of the respective active-sites, are marked with an open square. Note that the identity of the profiles is maintained up to residues -39 and 290, respectively. See text for explanation of negative sequence numbers. (From Martínez & Cid 1986.)

homology in the sequence and nearly a 100% conservation of the secondary structure, since the amino acid replacements have been made by hydrophobically equivalent residues.

Lys 274 in the pig kidney enzyme should correspond to Lys -54 in the rabbit enzyme, which has been reported to participate in the catalytic mechanism.

The location of the AMP binding site has been made by aligning the sequences 51 to 83 in the rabbit enzyme with 135 to 166 of the pig kidney enzyme, as seen in Fig. 8.10. Even though there is only a 50% conservation of the primary structure, the similarity between the profiles is evident. Lys 141 would be involved in AMP binding in the pig kidney enzyme, the analog to Lys 58 in the rabbit protein (Suda *et al.* 1982).

Domain I, in addition to the AMP binding site, contains the hyper-reactive SH group of Cys 128 (Chaterjee *et al.* 1984) and the region sensitive to proteolysis. This region would be the loop including residues 57 to 67 (Nakashima & Horecker 1971) which is well exposed in the model proposed. The 17-residue strand that joins both domains is flexible enough so as to allow their superposition in a way that permits the proximity of the AMP-binding site, the active-site and the hyper-reactive SH group (Fig. 8.11).

It must be emphasized here that the location of the active-site, and of the AMP site by secondary structure homology, agrees with that proposed by Marcus *et al.* (1982), based on primary structure homology found between pairs of hexapeptides in the pig kidney and rabbit liver enzymes.

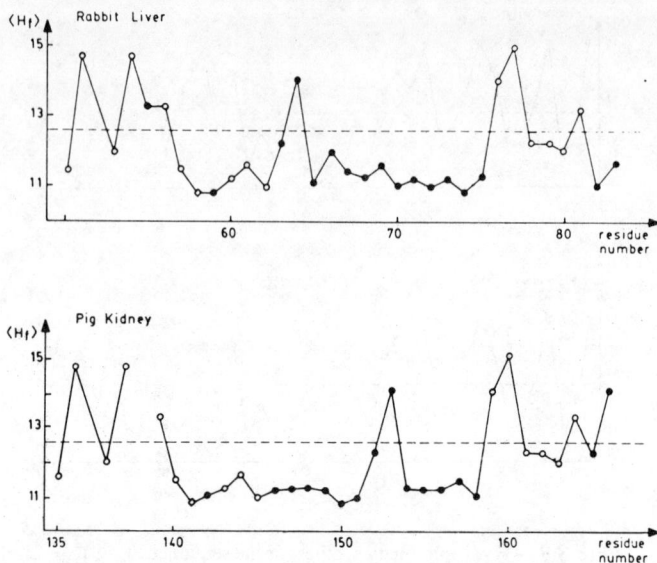

Fig. 8.10 — Hydrophobicity profiles for sequences 51 to 83 of rabbit liver and 135 to 166 of pig kidney fructose-1,6-bisphosphatases aligned for best fit. Note the similarity between the profiles, even though the conservation of the primary structure is low. Lys 58 and its analog Lys 141, involved in the AMP binding site, are marked with open squares. The nomenclature of Fig. 8.9 has been used. (From Martínez & Cid 1986.)

Fig. 8.12 illustrates the distances obtained in the proposed model of fructose-1,6-bisphosphatase between the AMP site, the active-site, and the hyper-reactive SH group. These distances agree with those obtained by ^{1}H-NMR and EPR studies on the divalent cation binding site of the rabbit liver enzyme (Cunningham *et al.* 1981). A similar distance has been found between the active-site and the hyper-reactive SH group by ^{1}H-NMR and ^{31}P-NMR studies of the bovine liver enzyme (Ganson & Fromm 1985).

Mechanism of inhibition by fructose-2,6-bisphosphate
The role played by fructose-2,6-bisphosphate in the inhibition of fructose-1,6-bisphosphatase is controversial. Some authors (Van Schaftingen & Hers 1981, Reyes *et al.* 1985) have reported the existence of an allosteric binding site for this metabolite, whereas others (Pilkis *et al.* 1981, Kitajima & Uyeda 1983) favor a direct interaction with the active-site, and consider it a competitive inhibitor. A third model (Corredor *et al.* 1984) postulates a biphasic behavior for fructose-2,6-bisphosphate, depending on the substrate concentration: low concentrations of fructose-1,6-bisphosphate would favor the interaction of fructose-2,6-bisphosphate with the active-site, whereas, at high concentration of the substrate the metabolite would bind to its own allosteric site, A biphasic behavior of the metabolite is consistent with the proposals of positions 1 and 2. This model can explain, for example, the

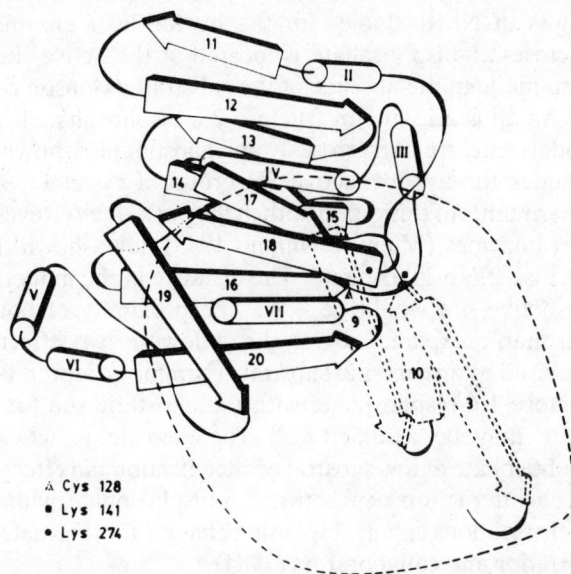

Fig. 8.11 — Superposition of the two domains in the proposed model structure of fructose-1,6-bisphosphatase from pig kidney. For the sake of simplicity only the contour of domain I and three important β-strands, β_{10} preceding Lys 141, β_9 containing Cys 128 and β_3 preceding the loop sensitive to proteolysis were represented. (From Martínez & Cid 1986.)

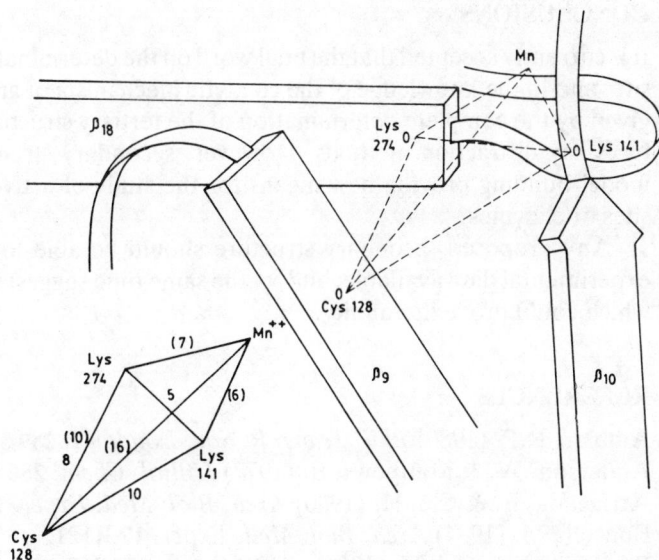

Fig. 8.12 — Distances measured in the proposed model correlated with those determined by Cunningham *et al.* (1981) by NMR and EPR studies in rabbit liver fructose-1,6-bisphosphatase. Cunningham's values are represented between brackets. The distances are in Å. (From Martínez & Cid 1986.)

results of NMR studies for the bovine liver enzyme which suggest that fructose-2,6-bisphosphate is located at the active-site, since these studies were made in the absence of the substrate (Ganson & Fromm 1985).

An allosteric site for fructose-2,6-bisphosphate is also postulated in the model structure of fructose-1,6-bisphosphatase shown in Fig. 8.12. This site includes the hyper-reactive SH group of cysteine 138, since modification experiments of this group both in the pig kidney (Reyes *et al*. 1985) and in rat liver enzymes (Meek & Nimmo 1983) have shown that it is protected by fructose-2,6-bisphosphate. The distance in the model between Cys 128 and the active-site would be 8 Å. The proximity of both sites would give a reasonable explanation for the following two effects: (1) the enzyme is inhibited by an excess of substrate (Nimmo & Tipton 1975); an interaction of fructose-1,6-bisphosphate with the allosteric site for fructose-2,6-bisphosphate may be assumed. (2) The enzyme is activated by fructose-2,6-bisphosphate at low substrate concentration, an effect that can be explained by an interaction of fructose-2,6-bisphosphate with the active-site. Both interpretations assume biphasic behavior for this metabolite as proposed by Corredor and collaborators (1984).

The three-dimensional model built according to the predicted secondary structure and to the chemical evidence available, allows us to propose reasonable locations for the active-site, the site sensitive to proteolysis, the allosteric site for AMP, and a possible allosteric site for fructose-2,6-bisphosphate, which can account for the kinetic behavior of the enzyme.

CONCLUSIONS

It is currently accepted that the final word on the determination of the active-site, and in the knowledge of the catalytic mechanism of an enzyme, will be given by the complete determination of the tertiary structure of the protein by X-ray diffraction methods. However, secondary structure studies and model building provide new means for the study of active and regulatory sites of enzymes.

Any proposed secondary structure should be able to account for the experimental data available, and, at the same time suggest new experiments which could prove its validity.

REFERENCES

Ambler, R.P. (1980) *Phil. Trans. R. Soc. London B* **289** 321–331.
Anderson, W. & Matthews, B. (1977) *J. Biol. Chem.* **252** 556–557.
Arriagada, E. & Cid, H. (1980) *Arch. Biol. Med. Exper.* **13** 44.
Bunster, M. (1984) *Arch. Biol. Med. Exper.* **17** R121.
Bunster, M. & Cid, H. (1984) *FEBS Letters* **175** 267–274.
Campos, M., Garcés, E., Montecinos, M., Ruiz, J. & Ward, P. (1979) *Arch. Biol. Med. Exper.* **12** 262.
Chatterjee, T., Edelstein, I., Marcus, F., Eby, J., Reardon, I. & Heinrikson, R. (1984) *J. Biol. Chem.* **259** 3834–3837.

Chou, P.Y. & Fasman, G.D. (1974) *Biochemistry* **13** 211–245.
Chou, P.Y. & Fasman, G.D. (1978) *Annual. Rev. Biochem.* **47** 251–276.
Cid, H., Bunster, M., Arriagada, E. & Campos, M. (1982) *FEBS Letters* **190** 247–254.
Citri, N. & Pollock, M.R. (1966), *Adv. Enzymol.* **28** 237–323.
Corredor, C., Boscá, L. & Sols, A. (1984) *FEBS Letters* **187** 199–202.
Cunningham, B.A., Raushel, F.M., Villafranca, J.J. & Benkovic, S.J. (1981) *Biochemistry* **20** 359–362.
Dale, J.W., Godwin, D., Mossakowaska, D., Stephenson, P. & Wall, S. (1985) *FEBS Letters* **191** 39–44.
Dijkstra, B., Kalk, K.H., Hol, W.G. & Drenth, J. (1981) *J. Mol. Biol.* **147** 97–122.
Dufton, P.Y. & Hider, R.C. (1977) *J. Mol. Biol.* **115** 177–193.
Dufton, P.Y. & Hider, R.C. (1983) *Eur. J. Biochem.* **173** 545–551.
Dufton, P.Y., Eaker, D. & Hider, R.C. (1983) *Eur. J. Biochem.* **137** 537–544.
Durkin, J.P. & Viswanatha, T. (1980) *Carlsberg Res. Commun.* **45** 411–421.
Fischer, W.K. & Thompson, E.D.P. (1983) *Aust. J. Biol. Sci.* **36** 235–250.
Ganson, N. & Fromm, H.J. (1985) *J. Biol. Chem.* **260** 2837–2843.
Kitajima, S. & Uyeda, K. (1983) *J. Biol. Chem.* **258** 7352–7357.
Knox, J., Kelly, J.A., Moews, P.C. & Murthy, N. (1976) *J. Mol. Biol.* **104** 865–875.
Levitt, M. & Chothia, C. (1976) *Nature (London)* **261** 552–558.
Lifson, S. & Sander, C. (1979) *Nature (London)* **282** 109–111.
Lim, V.I. (1974) *J. Mol. Biol.* **88** 857–872.
Marcus, F., Edelstein, I., Reardon, I. & Heinrikson, R. (1982) *Proc. Natl. Acad. Sci. USA* **79** 7161–7165.
Martínez, J. (1984) *Biochemist's thesis*, University of Concepción, Chile.
Martínez, J. & Cid, H. (1984) *Arch. Biol. Med. Exper.* **17** R156.
Martínez, J. & Cid, H. (1986) *Arch. Biol. Med. Exper.* **19** 77–83.
Matthew, M. (1978) *FEMS Microbiol. Lett.* **4** 241–244.
Meek, J.D. & Nimmo, H. (1983) *FEBS Letters* **160** 105–109.
Nakashima, K. & Horecker, B.L. (1971) *Arch. Biochem. Biophys.* **146** 153–160.
Nimmo, H. & Tipton, K. (1975) *Eur. J. Biochem.* **58** 567–574.
Nostrand, B. (1974), PhD. thesis, University of Uppsala, Sweden.
Pilkis, S.J., El Mahgrabi, M.R., McCrane, M.M., Pilkis, J. & Claus, T.H. (1981) *J. Biol. Chem.* **256** 11489–11495.
Ponnuswamy, P.K., Phrabhakaran, M. & Manavalan, P. (1980) *Biochim. Biophys. Acta* **623** 301–316.
Pontremoli, S. & Horecker, B.L. (1970) *The Enzymes* (3rd edn) **4** 611–646.
Pontremoli, S., Melloni, E. & Horecker, B.L. (1983) *Biochem. Soc. Trans.* **11** 241–244.
Reyes, A., Hubert, E. & Slebe, J.C. (1985) *Biochem. Biophys. Res. Commun.* **127** 373–379.
Seaton, B., Campbell, R., Petsko, G., Rose, D., Edelstein, I. & Marcus, F. (1984) *J. Biol. Chem.* **259** 8915–8916.

Soloway, B. & McPherson, A. (1978) *J. Biol. Chem.* **253** 2461-2462.

Suda, H., Xu, G., Kuthey, R., Natalini, P., Pontremoli, S. & Horecker, B.L. (1982) *Arch. Biochem. Biophys.* **217** 10–14.

Tsernoglou, D. & Petsko, A. (1976) *FEBS Letters* **68** 1–4.

Van Schaftingen, E. & Hers, H.G. (1981) *Proc. Natl. Acad. Sci. USA* **78** 2861–2863.

Waxman, D.J. & Strominger, J.L. (1980) *J. Biol. Chem.* **255** 3964–3976.

Walkinshaw, M.D., Saenger, W. & Maelicke, A. (1980) *Proc. Natl. Acad. Sci. USA* **77** 2400–2404.

Xu, G., Natalini, P., Suda, H., Tsolas, O., Dzugaj, A., Sun, S., Pontremoli, S. & Horecker, B.L. (1982) *Arch. Biochem. Biophys.* **214** 688–694.

9

Active-site studies by X-ray diffraction methods

Dr Hilda Cid, Laboratorio de Biofisica Molecular, Universidad de Concepción, Concepción, Chile

STATEMENT OF THE PROBLEM

An accurate knowledge of the mechanism of action of an enzyme could be obtained directly if the following two experiments could be performed:

(1) A direct observation of the enzyme molecule with enough resolution to appreciate all the details of the active site.
(2) A 'freezing' of the enzymatic reaction in order that all the intermediate enzyme–substrate and enzyme–product complexes last long enough to be identified.

The first experiment implies the determination of the tertiary structure of the enzyme, that is, the spatial location of every one or the thousands of atoms of its hundreds of amino acid residues. To discern an atom from its neighbors means to resolve distances of the order of 1 Å ($= 10^{-7}$ mm). Since the limit of resolution of the naked eye is 10^{-1} mm, a microscope capable of a magnification of $10^{6} \times$ would be needed. One can view, therefore, the determination of the tertiary structure of a protein (in fact, the structure of any molecule, large or small) as a problem of designing an adequate microscope.

Since enzymes are usually very efficient catalysts, the mean life of the intermediate complexes formed in the reaction process are too short to allow a direct observation, even with the most powerful microscope. Therefore the second experiment should be performed with poor substrates and under unfavorable conditions for catalysis. An alternative is to work with inhibitors which bind to the active site and remain there; in fact, several catalytic mechanisms have been solved by determining the exact location of inhibitors in the active site.

OPTICAL BASIS OF THE X-RAY DIFFRACTION METHOD

The resolution of a microscope cannot be better than the dimensions of the wavelength of the radiation used to 'illuminate' the object; therefore, in order to obtain a resolving power of 10^{-7} mm a visible-light microscope must be discarded. Wavelengths of 10^{-7} mm or lower in the electromagnetic spectrum belong to X-rays. According to De Broglie's theory, these waves are associated with electrons and protons. The most powerful microscopes actually in use are electron microscopes, but the best resolution obtained so far is not better than 10 to 20 Å, more than one order of magnitude larger than the resolution needed. The strong interaction between electrons and the sample observed plays against the possibility of improving this limit of resolution. A proton microscope shows even more disadvantages.

X-rays can easily penetrate samples, but have other drawbacks: the propagation of X-ray waves is not appreciably disturbed by any transparent media, nor by electric or magnetic fields. There are no lenses that can appreciably deviate this radiation, a necessary condition for image formation. The solution to this problem, and with it the discovery of the most powerful method to determine the tertiary structure of a protein, is based on optical principles established at the end of the last century by Abbe, and reformulated by Zernike in 1946 (Lipson 1972).

Abbe's theory states that the process of image formation takes place in two stages of diffraction. Here diffraction is understood in the more general form, that is, as a process that occurs naturally, every time the propagation of a wave is limited in some way.

In the microscope schematized in Fig. 9.1, the object is illuminated by

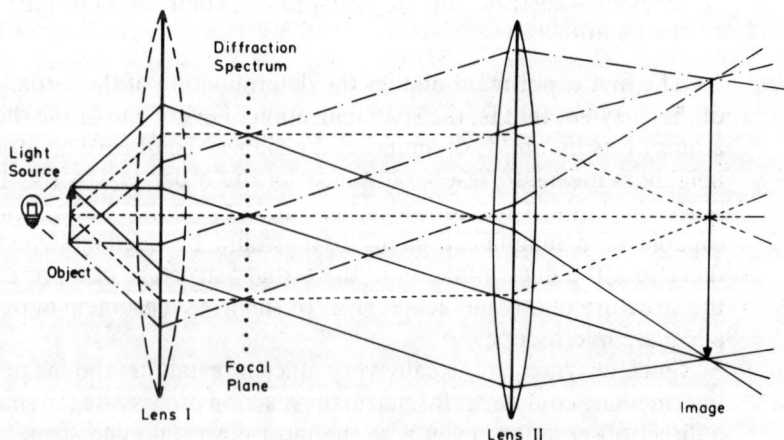

Fig. 9.1 — Abbe's theory of image formation in a microscope.

light coming from the source at the left, becoming a source of wavelets scattered in every direction of space. All wavelets scattered in a particular direction, originated from different points of the object will add to each other. The amplitude and the phase of the resulting wave will be determined by the amplitude and phases of the individual wavelets; thus the wave will depend on the structure of the object. The new light distribution produced by the diffraction phenomena is called the 'diffraction spectrum' of the object, and to a first approximation and for a planar object, it will be focused at the second focal plane of lens 1. It must be emphasized that the presence of this lens is not necessary to observe the diffraction spectrum, if the dimensions of the object are small enough to allow the proximity required for the combination of the wavelets originated on different points of the object.

In the second stage, all waves diverging from the diffraction spectrum are brought together by lens 2 in such a way that all waves originated at a single point of the object are focused at a single point at the image plane. This stage can be interpreted as a second diffraction, caused by the limitation of the wave front by lens 2. The light distribution at the image plane can be calculated mathematically with the aid of a computer, provided the amplitudes and phases of all the waves diverging from the diffraction spectrum are known.

When the light source is replaced by an X-ray source, the diffraction spectrum is produced without the presence of lens 1, and, if we can determine the amplitudes and phases of all the waves coming out from this diffraction spectrum, a computer can calculate the 'image' of our object. In this way one can have an 'X-ray microscope' without lenses and with a resolution power of the order of 10^{-7} mm.

MATHEMATICAL RELATIONSHIPS BETWEEN OBJECT, IMAGE, AND DIFFRACTION SPECTRUM

The amplitudes and phases of a scattered wave in a certain direction will be the result of the sum of all the wavelets scattered in that direction by all the points of the object. If we represent by $F(\alpha)$ the wave scattered in a direction making an angle α with the optical axis of the microscope (Fig. 9.2), for a one-dimensional object we can write

$$F(\alpha) = \int_{-\infty}^{\infty} f(x) \exp\left[\frac{i2\pi x \sin \alpha}{\lambda}\right] dx \qquad (1)$$

where $f(x)$ represents the amplitude of the wavelet scattered in the direction α by a point located at the distance x from the axis. Making $\sin \alpha/\lambda = X$ the expression simplifies to

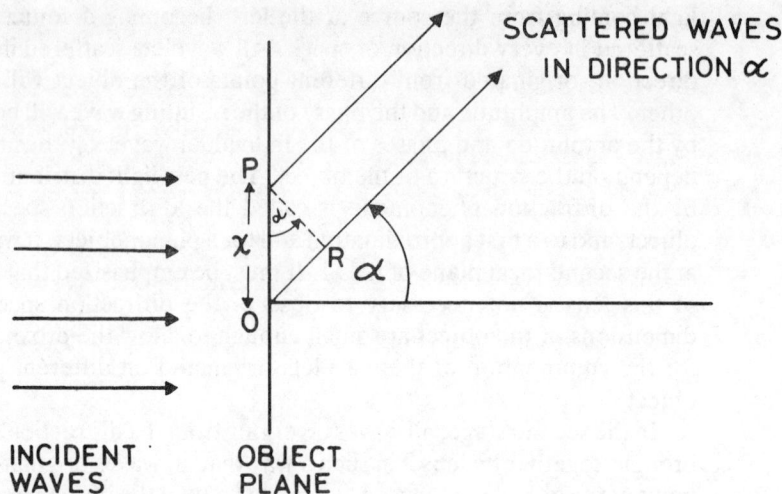

Fig. 9.2 — Calculation of the phase difference for two wavelets diffracted in the direction α, originated at the origin and at a point P located at a distance *x* from the origin. OR is the optical path difference.

$$F(X) = \int_{-\infty}^{\infty} f(x)\, \exp(2\pi i Xx)\, \mathrm{d}x \qquad (2)$$

which is the expression of a Fourier Transform. This formula states that $F(X)$ is the Fourier Transform of $f(x)$. $F(x)$ represents the 'light' distribution at the diffraction spectrum, $|F|2 = FF^*$ is the intensity of the diffracted wave. The meaning of $f(x)$ can be better understood if we identify our linear object with a linear molecule of several atoms located at distances x_j from the optical axis. To a first approximation we can consider discrete centers of scattered wavelets, and, since the electrons are responsible for the scattering of the X-rays, the intensity of each scattered wavelet will be a direct function of the number of electrons of the scattering atom. The continuous function $f(x)$ becomes the 'scattering power' of each atom j, a function of the electron distribution in the object. The expression of the Fourier Transform, considering point atoms, is changed to

$$F(X) = \sum_j f_j \exp(2\pi i Xx_j) \qquad (3)$$

$F(X)$ is called the 'structure factor' since it depends on the location (x_j) and on the nature f_j of the atoms, that is, the structure of the object. The structure factor is now given as a function of a coordinate X in the plane of the diffraction spectrum, instead of the variable α.

By an analogous deduction, it can be shown that the mathematical relationship between the image (I) and the diffraction spectrum is

$$I(x) = \int_{-\infty}^{\infty} F(X) \exp(-2\pi iXx) \, \mathrm{d}X \tag{4}$$

In words, the image is the Fourier Transform of the Fourier Transform of the object.

For a three-dimensional object, one must deal with a three-dimensional image and a three-dimensional diffraction spectrum (equations 5 and 6).

Even though object and image could be assumed to be discontinuous functions, the diffraction spectrum is a continuous function:

$$I(xyz) = \int_{-\infty}^{\infty} \int_{-\infty}^{\infty} \int_{-\infty}^{\infty} F(XYZ) \exp[-2\pi i(Xx + Yy + Zz)] \, \mathrm{d}X\mathrm{d}Y\mathrm{d}Z \tag{5}$$

$$F(XYZ) = \sum_{j} f_j \exp[2\pi i(x_j X + y_j Y + z_j Z)] \tag{6}$$

For a detailed treatment of the use of Fourier methods in crystallography see Ramachandran and Srinivasan (1970).

DIFFRACTION SPECTRUM OF A REPEATING OBJECT

In order to calculate the image, it is necessary to know as precisely as possible, the amplitudes and phases of all the waves emerging from the diffraction spectrum. The amplitude of any wave is proportional to the square root of its intensity. Therefore, the first problem is solved by a careful measurement of the intensity distribution at the diffraction spectrum. However, a single molecule, even a protein molecule with thousands of atoms, is unable to produce diffracted X-rays strong enough to be detectable. It is necessary to obtain the cooperative diffraction of several thousands of such molecules to be able to measure, on a film, or by means of a radiation detector, the diffracted X-ray beams. This cooperative diffraction is obtained with an ordered arrangement of the molecules we want to observe. A perfect three-dimensional periodic repetition of a molecule is a crystal, and the three vectors defining the repetition periods form the 'unit cell' of the crystal. Therefore, the first problem to be solved in the determination of the tertiary structure of a protein is to obtain protein crystals.

What is the relationship between the diffraction spectra produced by a single molecule and that produced by the periodic repetition of the same molecule?

Fig. 9.3 shows the diffraction spectrum of a circular hole and that

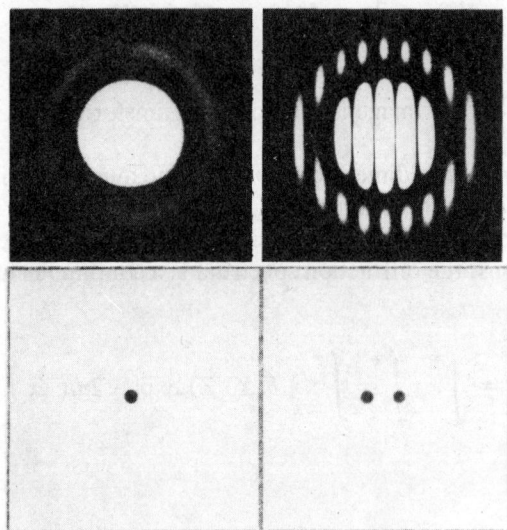

Fig. 9.3 — The effect of repeating the object in the diffraction spectrum. (From Harburn *et al*. 1975.)

produced by two identical holes. It is clear that the diffraction spectrum of the two holes has the same intensity distribution as the spectrum produced by a single aperture, except that it is crossed by dark, equally spaced, straight interference lines, perpendicular to the translation vector relating both circular apertures. The repetition of the diffracting object did not generate new bright zones; instead, it 'selected' some regions within the bright zones produced by the diffraction of a single object. Since the amplitudes of two identical waves are added in the bright zone, the intensity is four times that obtained with a single aperture. This is the classical Young experiment; it can be shown that the spacing between the dark interference lines is inversely related to the length of the vector relating both circular holes.

If the object is repeated in two dimensions, the diffraction spectrum will be crossed by two sets of interference lines, with orientations and spacings depending on the direction and magnitudes of the translation vectors. A three-dimensional repetition of the object will produce a more strict selection of the remaining bright zones, reducing then to a collection of bright spots.

Good examples of the effects of repetition on the diffraction spectra can be found on the book by Haburn *et al.* (1975).

Mathematically, a crystal can be considered as a convolution of the repeating molecules and a lattice (the 'real lattice') formed by the repeating vectors. The diffraction spectrum of the crystal, or Fourier transform, is the product of the diffraction spectrum of the single molecules and the interference pattern of the real lattice. This interference pattern is also a lattice, called the 'reciprocal lattice'. Since a lattice is a function with non-zero values only at the lattice nodes, the product of the diffraction spectrum of the single molecule and the reciprocal lattice will have non-zero values only at the reciprocal lattice nodes.

The diffraction spectrum of a crystal is no longer a continuous function of three coordinates XYZ, but it will depend on three integers (hkl) that define the nodes of the reciprocal lattice:

$$F(hkl) = \sum_j f_j \ \exp[2\pi i(hx_j + hy_j + lz_j)] \tag{7}$$

This formula allows a calculation of the diffraction spectrum of a molecule, provided the nature and position of all its j atoms are known. It is very useful to test the reliability of the structure determined: if correct, the observed structure factors must be similar to the calculated ones.

Fig. 9.4 shows a plane of the diffraction spectrum of a protein crystal. As expected, it consists of discrete points very close to each other. Since the molecular dimensions require large repetition vectors, the reciprocal vectors are small.

THE ELECTRON DENSITY FUNCTION

Up to now we have considered a protein molecule as a collection of point atoms with all the electrons concentrated at points defining the location of these atoms. An alternative and more realistic way to define the structure of a protein is in terms of the electrons contained in the unit cell, assuming a continuous distribution of electrons. A continuous function called 'electron density' is defined and represented by $\rho(xyz)$, where $\rho(xyz)$ dxdydz represents the number of electrons included in the volume dxdydz at the point (xyz) of the unit cell. As we have pointed out before, the electrons are responsible for the scattering of the X-rays, therefore the X-ray diffraction spectrum is a function of the electron distribution.

The mathematical expression (5) used to calculate the three-dimensional image becomes:

$$\rho(xyz) = \frac{1}{V} \sum_h \sum_k \sum_l \ F(hkl) \ \exp[-2\pi i(hx + ky + lz)] \tag{8}$$

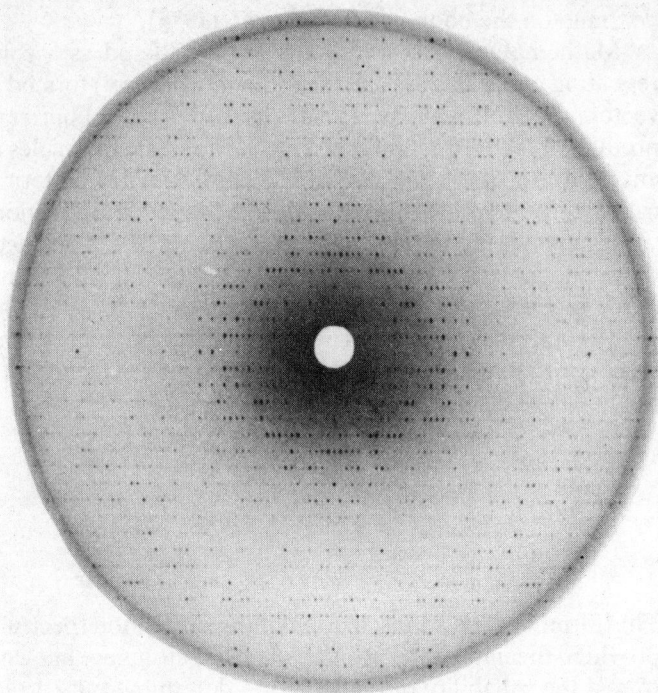

Fig. 9.4 — A typical diffraction spectrum from a protein crystal. Precession
photograph of the (hkO) plane of the actin-profilin complex.

considering that *hkl* are integers and not continuous variables.

The structure factor $F(hkl)$ represents the diffracted wave at the point
(hkl) of the diffraction spectrum. This wave has an amplitude and a phase.
The amplitude is the modulus of the complex quantity $F(hkl)$, and the phase
can be represented as $\alpha(hkl)$; this can be written as:

$$F(hkl) = |F(hkl)| \exp[i\alpha(hkl)] \qquad (9)$$

and the electron density function becomes

$$\rho(xyz) = \frac{1}{V} \sum_h \sum_k \sum_l |F(hkl)|$$
$$\times \exp[-2\pi i(hx + ky + lz) + \alpha(hkl)] \qquad (10)$$

The amplitude $F(hkl)$ can be measured as the square root of the intensity of each diffracted spot (*hkl*) in a diffraction spectrum such as that shown in Fig. 9.4. The determination of the phase angle $\alpha(hkl)$ is the second key problem that protein crystallographers need to solve. This is accomplished by 'marking' the protein with heavy atoms, that do not disturb the packing of the native protein molecule (Blundell & Johnson 1976). Theoretically, a minimum of two different such heavy-atom derivatives need to be obtained and measured, in the same way as for the native protein. In fact, every protein structure solved utilizes four or more such derivatives. Only when this task is successfully accomplished, can one be sure that the three-dimensional structure of the protein molecule can be solved.

Formula 10 is used by the computer to calculate the 'image', in every point (*xyz*) of the unit cell, as requested by the programmer. Usually one divides the unit cell axes in 50 (or 100) points, and asks the computer to calculate $\rho(xyz)$ for all values of *x* from 0/50 to 50/50, and the same for the variables *y* and *z*. The computer calculates the value and plots a number in the corresponding point of the unit cell. Usually it also contours the electron density map, by joining all the points with the same value for the electron density function. For a three-dimensional electron density function, all calculations are performed in sections, with one coordinate held constant on each section.

Fig. 9.5 shows the calculated electron density function for a small planar

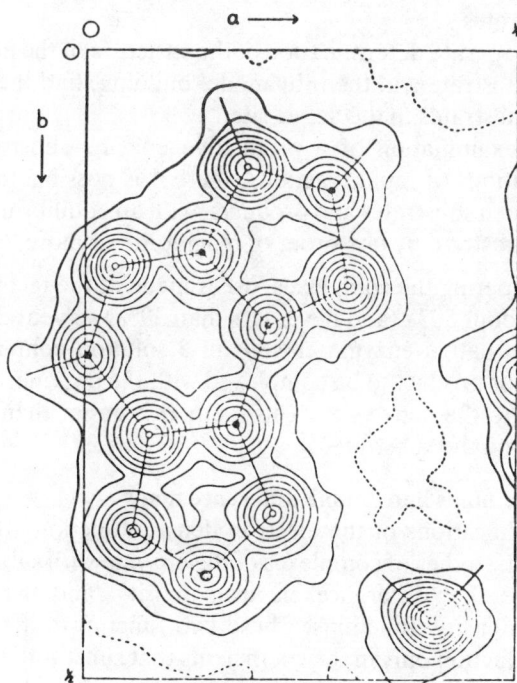

Fig. 9.5 — Electron density map of a small planar molecule. The height of the peaks, given by the number of contours, is proportional to the number of electrons of the atom defined by that peak.

molecule; the curves joining points of equal electron density define the position of the atoms. Each atom is clearly resolved from its neighbors: the 'X-ray microscope' has reached a magnification power of $10^6 \times$.

ACTIVE SITE AND CATALYTIC MECHANISM

The most remarkable characteristic of enzyme catalysis is that the enzymes bind their substrates at the active site in a unique orientation that allows the desired proximity of all the necessary elements that take part in the reaction. To understand the mechanism of an enzyme it is necessary to know the structure of the native enzyme, and that of the complexes of the enzyme with its substrates, intermediates and products as well (Fersht 1985). As pointed out before, usually the mean life of enzyme–substrate complexes can be of the order of seconds or less, and the recording of the X-ray data takes several hours.

To overcome these difficulties several strategies are used. Some of them are:

(1) The determination of the structure of an enzyme-inhibitor, enzyme-product or enzyme–substrate analog complex.
(2) The determination of the structure of the enzyme–substrate complex under unfavorable reaction conditions, such as very low temperature, a pH where the substrate is in the wrong ionic state, or to use very poor substrates.
(3) The separate determination of the structure of the native enzyme and of the substrate and then, by model building, find the correct location of the substrates in the active site.
(4) The examination of a productive enzyme–substrate complex under conditions of rapid reaction, where it is possible to set an equilibrium between substrates and products, with an equilibrium constant favoring the substrate by one order of magnitude or more.

In preparing the complexes, use is made of the fact that protein crystals contain about 50% (and never less than 30%) solvent. It is thus possible to 'soak' the native enzyme crystals in a solution containing the substrate, inhibitor or product to be complexed with the enzyme. The ligand will then diffuse into the active site. This method is easier than co-crystallizing the enzyme and the substrate.

Difference maps and double difference maps

Two modifications of the electron density function are used in the X-ray diffraction studies of complexes of the enzyme with substrates or inhibitors. These are the 'difference electron density' and the 'double difference electron density' functions. These two functions differ from the electron density function only in the coefficients that substitute the absolute value of the structure factor in equation (10), as shown in Table 1.

The meaning of these two functions is shown schematically in Fig. 9.6 for the special case of a linear molecule. It is assumed that $\rho_1(x)$ represents the structure (or electron density) of the native enzyme which has three atoms at

Table 1

Function	Coefficient	Calculation performed
Electron density	$\|F(hkl)\|$	$\rho(xyz)$
Difference Electron Density	$\|F_2(hkl)\| - \|F_1(hkl)\|$	$\rho_2(xyz) - \rho_1(xyz)$
Double Difference Electron density	$2\|F_2(hkl)\| - \|F_1(hkl)\|$	$2\rho_2(xyz) - \rho_1(xyz)$

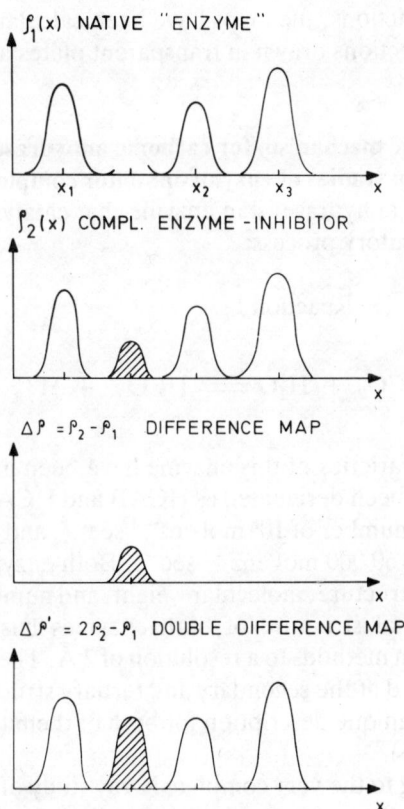

Fig. 9.6 — Difference and double difference electron density maps illustrated for a one-dimensional 'molecule'.

locations x_1, x_2 and x_3; $\rho_2(x)$ is the structure of the enzyme-inhibitor complex with the inhibitor located at position x_4. The difference electron density function shows only one peak at position x_4, thus indicating the location of the inhibitor. The double difference function shows all the maxima corresponding to the native 'protein', plus a double weighted peak indicating the position of the inhibitor. The importance of the double difference map is that it displays the relative position of the inhibitor with respect to other residues involved in the active site.

The use of these two functions is restricted to the condition that there are no drastic changes in the structure or in the packing of the enzyme on binding of the ligand. The calculation is then based on the intensity differences obtained from the diffraction spectra of the complexed and native enzymes and on the phase angles obtained from the native enzyme.

Fig. 9.7 shows electron density maps together with the corresponding difference and double difference maps for the enzyme carbonic anhydrase B, native and complexed with Diamox, a sulfonamide noncompetitive inhibitor. Since in this case all the Fourier summations are three-dimensional functions, the maps have been calculated in sections. Fig. 9.7 shows several sections drawn in transparent plates and stacked together.

A catalytic mechanism for carbonic anhydrase obtained by X-ray diffraction studies of enzyme-inhibitor complexes

Carbonic anhydrase is an enzyme that catalyses a fundamental reaction in the respiratory process:

Reaction I

$$CO_2 + H_2O \rightleftharpoons HCO_3^- + H^+$$

Two varieties of this enzyme have been isolated from human red cells, and have been designated as HCAB and HCAC, respectively. HCAC has a turnover number of 10^6 mol enz^{-1} sec^{-1}, and HCAB, which is 'slower', has a rate of 160 000 mol enz^{-1} sec^{-1}. Both enzymes show great similarities in primary structure, molecular weight, and number of amino acids. The three-dimensional structure of both enzymes has been determined by X-ray diffraction methods to a resolution of 2 Å. The similarity of these enzymes is maintained at the secondary and tertiary structure level so that it is possible to give a unique description for both of them (Kannan *et al.* 1975, Nostrand *et al.* 1975).

Owing to the very complete X-ray study of the complexes of HCAB and HCAC with inhibitors, this enzyme is a good example of the information that X-ray diffraction methods can provide on the catalytic mechanism of an enzyme.

The secondary structure of carbonic anhydrase HBAB and HCAC is shown in Fig. 9.8 in the 'cylinder and arrow' representation. The Zn^{+2} ion, a cofactor of the enzyme, is coordinated to three histidine residues. Both the location of the Zn ion and its coordination to three histidines and to a solvent molecule in a distorted tetrahedron were determined from the electron density maps. The Zn peak, with its 30 electrons, gave by far the highest

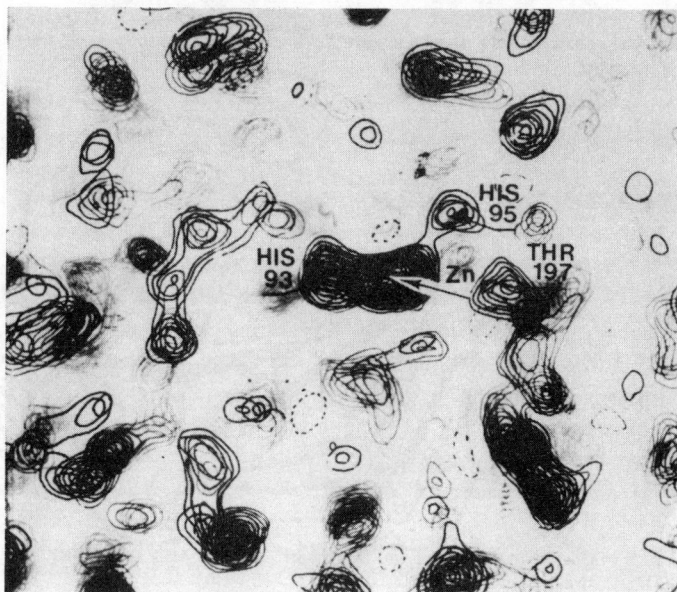

(a)

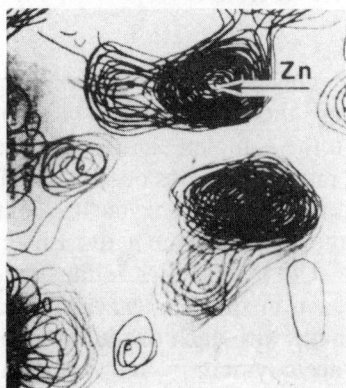

(b)

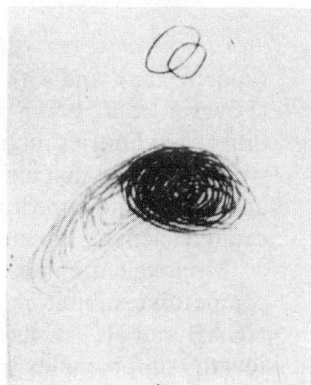

(c)

Fig. 9.7 — (a) Electron density maps of the active site region in native HCAB. (b) Double difference map for the complex HCAB-Diamox. Note that the shape of the maximum including the Zn atoms is the same as in (a). (c) Difference map between the complex HCAB-Diamox and the native enzyme. Only the peaks corresponding to the Diamox inhibitor are present. (From Nostrand 1974.)

maximum density in the map. The non-histidine fourth ligand is a water molecule or an OH^-.

Once the tertiary structure of these enzymes was solved, it was possible to locate the active site and to determine the role of the metal atom. As part of the active site studies, the enzyme was prepared without the Zn ion

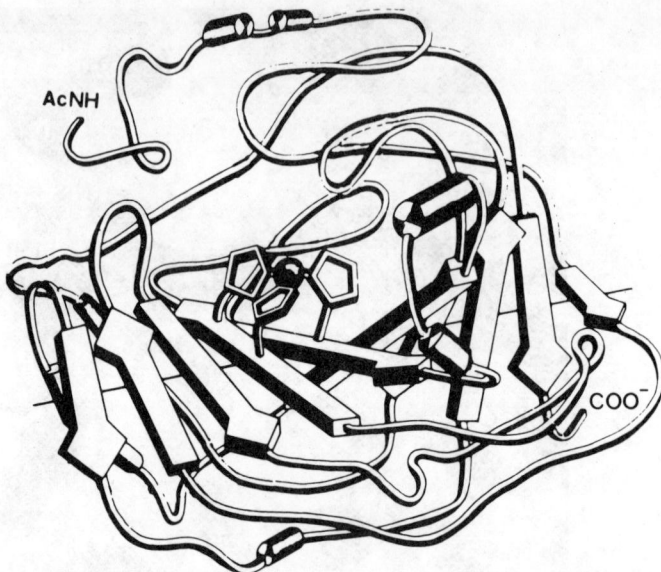

Fig. 9.8 — Secondary structure of HCAB and HCAC in the cylinder-arrow representation. The Zn atom is shown attached to the three histidyl ligands. (From Nostrand 1974.)

(apoenzyme), and with the Zn^{+2} replaced by Mn^{+2}, Co^{+2}, Cu^{+2} or Mg^{+2} (Lövgren *et al*. 1971). In each case the metal position was obtained from difference Fourier maps between the complex and the apoenzyme. These studies showed that although all these metal ions occupy the same location and bind to the same ligands as Zn^{+2}, only the enzyme–Co complex retains catalytic activity (about 50% of the activity of the native Zn–enzyme).

Monovalent anions (I^-, Br^-, Cl^-), as well as sulfonamides, are non-competitive inhibitors of the reaction of both enzymes. Complexes of HCAB and HCAC have been made with each of the three ions and with several sulfonamides such as 3-acetoxymercury-4-aminobenzene sulfonamide (AMsulf), acetazolamide, Diamox and salamide (Bergstén *et al*. 1972). All of these complexes were studied by X-ray diffraction methods, and the position of the inhibitors were located by difference and double difference Fourier maps, such as those shown in Fig. 9.7.

The results of these studies have clearly shown that:

(1) The monovalent anions bind to the Zn ion by replacing the fourth ligand. The electron density peak corresponding to the anion was found in the location occupied by water, or OH^- in the native enzyme. Fig. 9.9 is a computer drawing of the I^- coordination at the active site of HCAC.
(2) All sulfonamide complexes bind to the Zn ion through the nitrogen of the sulfonamic group. The nitrogen atom also replaces the fourth Zn ligand. Some of the amino acid residues present in the active site also bind to the sulfonamide inhibitors as seen in Figs 9.10 and 9.11.

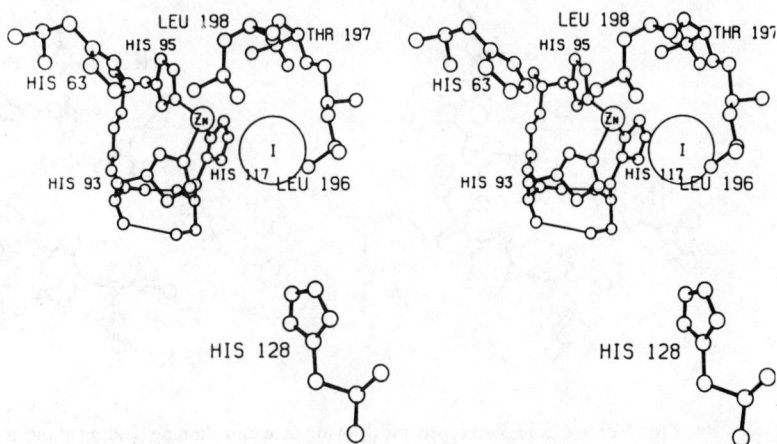

Fig. 9.9 — Location of I$^-$ in the active site of HCAC. Stereoscopic (ORTEP) drawing. (From Bergstén et al. 1972.)

The only known competitive inhibitor for the hydration reaction of CO_2 catalysed by carbonic anhydrases is imidazole. This acts only with HCAB. Even though this inhibitor is a small molecule, its shape made its location within the active site possible. Electron density maps of the enzyme–imidazole complex, and difference Fourier maps between the complex and the native enzyme, showed that this inhibitor binds to the Zn-ion without disturbing the fourth ligand. It is attached to Zn at a fifth coordination position, with the nitrogen of the ring 2.7 Å from the Zn ion (Kannan et al. 1977).

This observation suggests that the substrate CO_2 can also occupy this fifth ligand position; this can put CO_2 in a highly favorable orientation for a reaction with the fourth Zn ligand, water.

Once the existence of a fifth coordination position for the Zn had been established, the rest of the experimental results began to fit. This explains, for example, why the Co–enzyme retains activity, since Co is the only other divalent cation with 5-fold coordination.

It was also shown that, although the N atom in sulfonamide replaces the fourth ligand, one of the oxygens of the sulfonamide group occupies the fifth coordination position of the Zn ion. This fact agrees with reports that sulfonamides act as competitive inhibitors of the hydration reaction of CO_2 by HCAB (Lindskog et al. 1971).

The proposed catalytic mechanism for carbonic anhydrases is shown in Figs 9.12 and 9.13. Two proposals have been made, since the data available were not sufficient to decide if the fourth Zn ligand was a water molecule or an OH$^-$ ion. In both mechanisms, the amino acid residues Glu 106 and Thr 199 take part. Both residues are conserved in all primate and ruminant carbonic anhydrases.

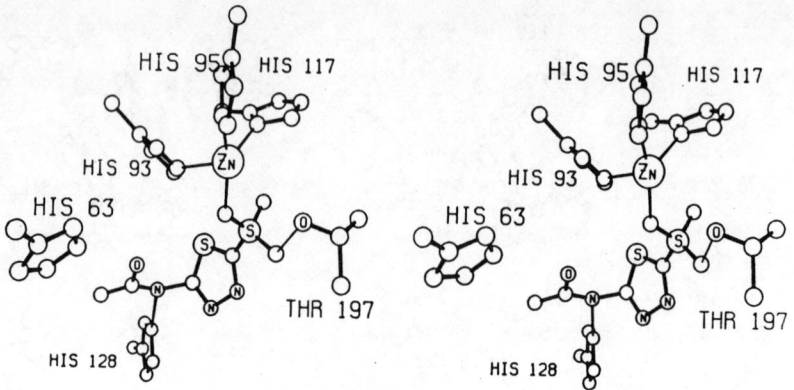

Fig. 9.10 — ORTEP stereoscopic drawing of acetazolamide binding to the active site of HCAC. (From Bergstén *et al*. 1972.)

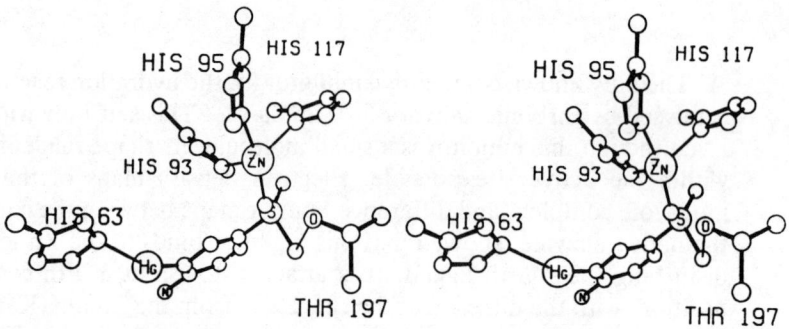

Fig. 9.11 — ORTEP stereoscopic drawing of the AMsulf molecule binding at the active site of HCAC (From Bergstén *et al*. 1972.)

X-ray diffraction studies of productive enzyme–substrate complexes under appropriate equilibrium conditions. Active site studies of triose phosphate isomerase

Triose phosphate isomerase, a dimeric enzyme with 247 amino acid residues in each of the identical chains, catalyses the following reaction:

Reaction II

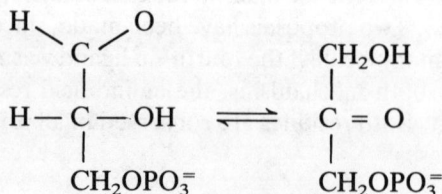

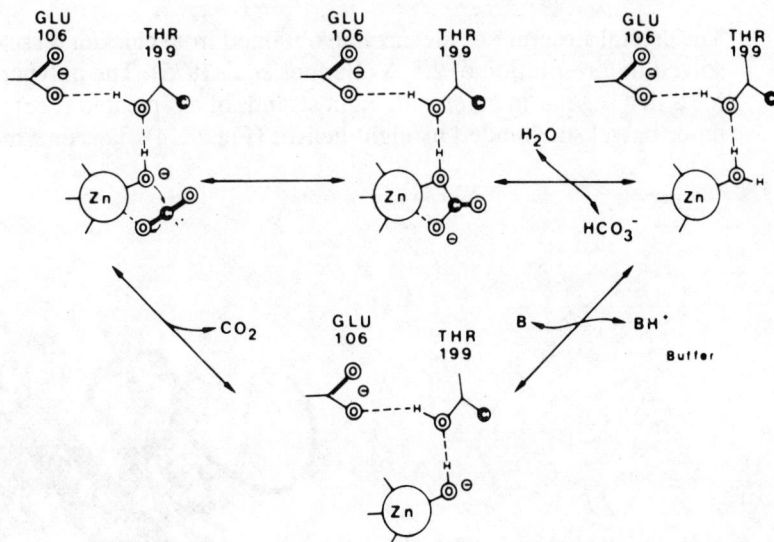

Fig. 9.12 — Proposed catalytic mechanism for carbonic anhydrase B, assuming that the fourth Zn ligand is OH⁻. (From Kannan *et al.* 1977.)

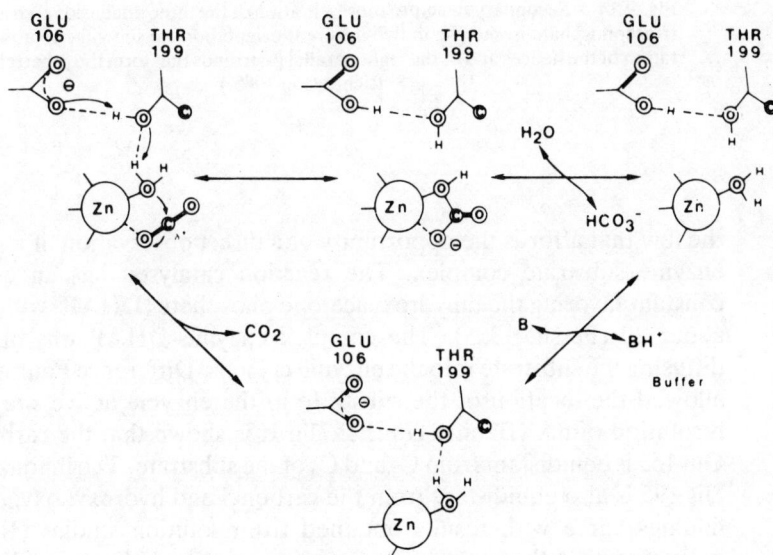

Fig. 9.13 — Proposed catalytic mechanism for HCAB, assuming that the fourth Zn ligand is a water molecule. (From Kannan *et al.* 1977.)

The crystal structure of the enzyme obtained from chicken muscle has been solved to a resolution of 2.5 Å (Banner *et al*. 1975). The polypeptide chain folds in a regular manner with eight strands of a β-pleated sheet forming an inner barrel surrounded by eight helices (Fig. 9.14). This enzyme is one of

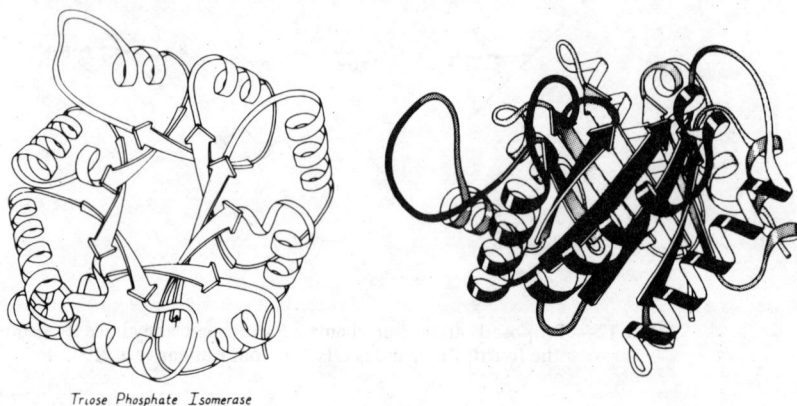

Triose Phosphate Isomerase

Fig. 9.14 — Secondary structure representation of the three dimensional structure of triosephosphate isomerase in two different orientations to show the relative orientation of the helices and of the eight parallel β-strrands that form the β-barrel. (From Richardson 1985.)

the few that affords the opportunity of a direct observation of a productive enzyme–substrate complex. The reaction catalysed has an equilibrium constant favoring the dihydroxyacetone phosphate (DHAP) with a value of about 20 (Fersht 1985). The complex enzyme–DHAP was obtained by diffusing the substrate into the enzyme crystals. Difference Fourier methods allowed the location of the substrate in the enzyme active site even at a resolution of 6Å (Banner *et al*. 1971). It is shown that the carboxylate of Glu 165 is equidistant from C_3 and C_4 of the substrate. The imidazole ring of His 195 is also equidistant from the carbonyl and hydroxyl oxygens. These findings agree with results obtained from solution studies (Rose 1975) indicating that the reaction occurs via a cis enediol intermediate with a proton transferred by a single base: Glu 166 (Fig. 9.15). The proposed active site lies at the carboxyl end of the β-barrel, and it is formed by residues from β-strands a, e, f, g, h and from helices D_1, E_1, and H_1, with participation of a few residues of the adjacent subunit (Fig. 9.14).

A

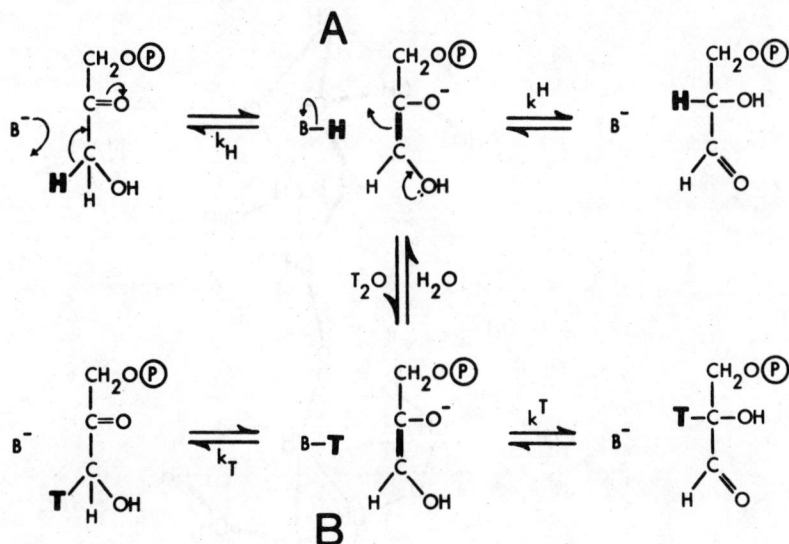

B

Fig. 9.15 — The enediol mechanism proposed by Rose for the reaction catalysed by triosephosphate isomerase. A base on the enzyme (B) removes the α-proton to give an enzyme–bound enediolate. The intermediate picks up a proton to form the product. The lower half of the figure illustrates the experiment made with tritiated water that led to this proposed mechanism. (From Knowles *et al*. 1972.)

A combination of X-ray diffraction methods and model building in the active site studies of lysozyme

Lysozyme catalyses the hydrolysis of a polysaccharide that is a major constituent of the cell wall of some bacteria. The substrate is formed by alternating units of N-acetylglucosamine (NAG) and N-acetylmuramic acid (NAM) (Fig. 9.16). The three-dimensional structure of this enzyme was totally solved by 1965 (Blake *et al*. 1965) and it has been further refined to a resolution of 1.5 Å, becoming one of the best detailed protein structures (Artymiuk & Blake 1981).

The enzyme has a molecular weight of 14 600 with 120 amino acid residues in a single polypeptide chain and four disulfide bridges. Lysozyme cleaves the polysaccharide chain in a specific position, between a C_1 of NAM and the oxygen attached to the C_4 of NAG (Fig. 9.16).

The enzyme has a cleft on one side. This cleft presents regions with amino acid residues which facilitate the binding of the non-polar regions of the substrate and also hydrogen bonding sites for the acylamino and hydroxyl groups. The crevice is divided into six sites designated as ABCDEF. NAM residues can bind to B, D, and F only, while NAG residues can occupy any site. The bond to be cleaved must lie between sites D and E. Fig. 9.17 illustrates the location of several saccharide inhibitors of lysozyme, as obtained from low resolution difference electron density maps.

The structure of the enzyme complexed with the trimer NAG-NAG-

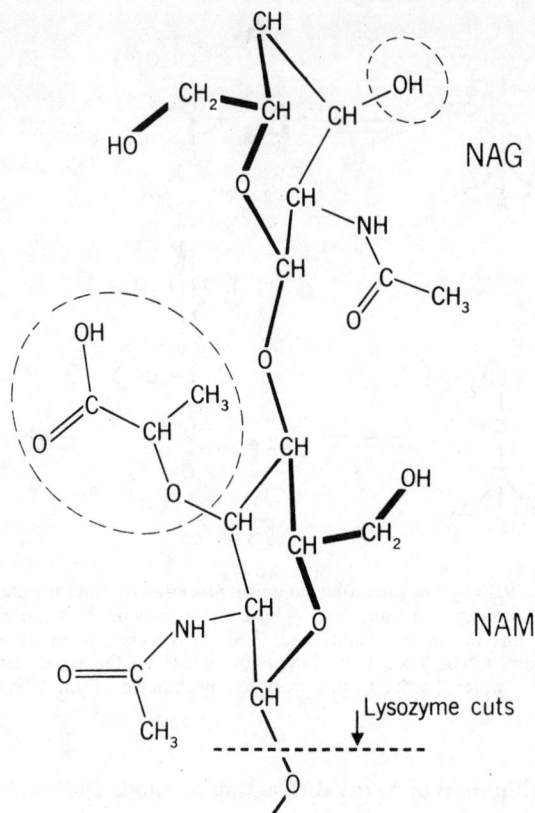

Fig. 9.16 — One unit of the (NAG-NAM)$_n$ substrate of lysozyme, showing the cleaved bond. (From Dickerson & Geis 1969.)

NAG has also been solved (Phillips 1967). This trimer behaves as a stable inhibitor, whereas longer NAG polymers are substrates, the velocity of cleavage being a direct function of the number of sugar rings up to hexasaccharides. Fig. 9.18 shows the α-carbon representation of the structure of lysozyme with the trisaccharide (NAG)$_3$ located at the active site.

(NAG)$_3$ proved to be a very poor substrate. Fourier maps showed that it binds to sites ABC and avoids the sites where the cleavage takes place. The structure of a productive substrate–enzyme complex was obtained by building a wire model of the complex enzyme-(NAG)$_3$ and extending the polysaccharide chain by adding more NAG units. Chemical intuition was used to select appropriate contacts between the enzyme's three-dimensional model and the polysaccharide chain. These studies determined that the bond to be cleaved is placed between the carboxyl groups of Glu 35 and Asp 52 (Phillips 1967).

Model building is currently done using a computer program to optimize the fitting between the enzyme and the substrate. Interactive graphic systems display the atomic coordinates of the enzyme and of the substrate in a television screen and the program allows selective rotations or displacements in order to obtain the best fit. This procedure has been used in a combination of X-ray studies and model building of the complex of lysozyme and the trisaccharide NAM-NAG-NAM (Fig. 9.19a). The crystals of the enzyme and the trisaccharide were grown by co-crystallization (Kelly *et al.* 1979). The coordinates of the trisaccharide were obtained by docking the atomic coordinates of two crystal structures β-(1-4)-N,N'-diacetylchitobiose (monohydrate) and N-acetyl-α-D-muramic acid. The fitting of the substrate in the difference electron density maps was made with an interactive graphics system and it is shown in Fig. 9.19b. A list of close contact distances (less than 5 Å between the trisaccharide and the native enzyme) clearly show the participation of Asn 103 and Asp 101 in the B site; Ala 107, Trp 108. Gln 57, Trp 63 and Asn 59 in the C site, and Asp 52, Val 109, Asn 46, and Glu 35 in the D site.

A different approach to the model building process has been presented by Pincus *et al.* (1976, 1977). Successive NAG residues were added to the rigid active site of lysozyme, and the dihedral angles φ and ψ, together with six rigid body variables that describe the orientation of the oligosaccharide substrates were allowed to vary to give unique low energy structures. This method was used with the complexes lysozyme-$(NAG)_4$, lysozyme-$(NAG)_5$ and lysozyme-$(NAG)_6$. Even though the results are promising, it was clear to the authors that the restriction of a rigid active site can introduce important distortions in the calculated energies. In any case, the model building performed after the energy minimization process allowed the removal of several unfavorable contacts between the polysaccharide and the enzyme (Fig. 9.20).

Active site studies in serine-proteases. Examples of X-ray diffraction studies on an acyl–enzyme intermediate, an enzyme–product complex and on a complex with a naturally occurring polypeptide inhibitor.
Chymotrypsin, trypsin, and elastase are three pancreatic enzymes that are structurally and kinetically very similar, hydrolysing peptides and synthetic ester substrates. Their primary structures are composed of approximately 50% identical or similar amino acid residues, and the similarity in their tertiary structures allows a unique structural description (Fersht 1985). Specificity is the major difference between these enzymes. Trypsin is specific for the positively charged amino acids lysine and arginine; chymotrypsin acts on the large hydrophobic chains of phenylalanine, tyrosine, and tryptophan, whereas elastase is effective with small hydrophobic amino acids such as alanine (Fig. 9.21).

When the tertiary structures of α-chymotrypsin (Matthews *et al.* 1967), elastase (Shotton & Watson 1970) and bovine trypsin (Huber *et al.* 1974, Stroud *et al.* 1974, Bode and Schwager 1975) were determined by X-ray diffraction methods, it was shown that the polypeptide backbones of these

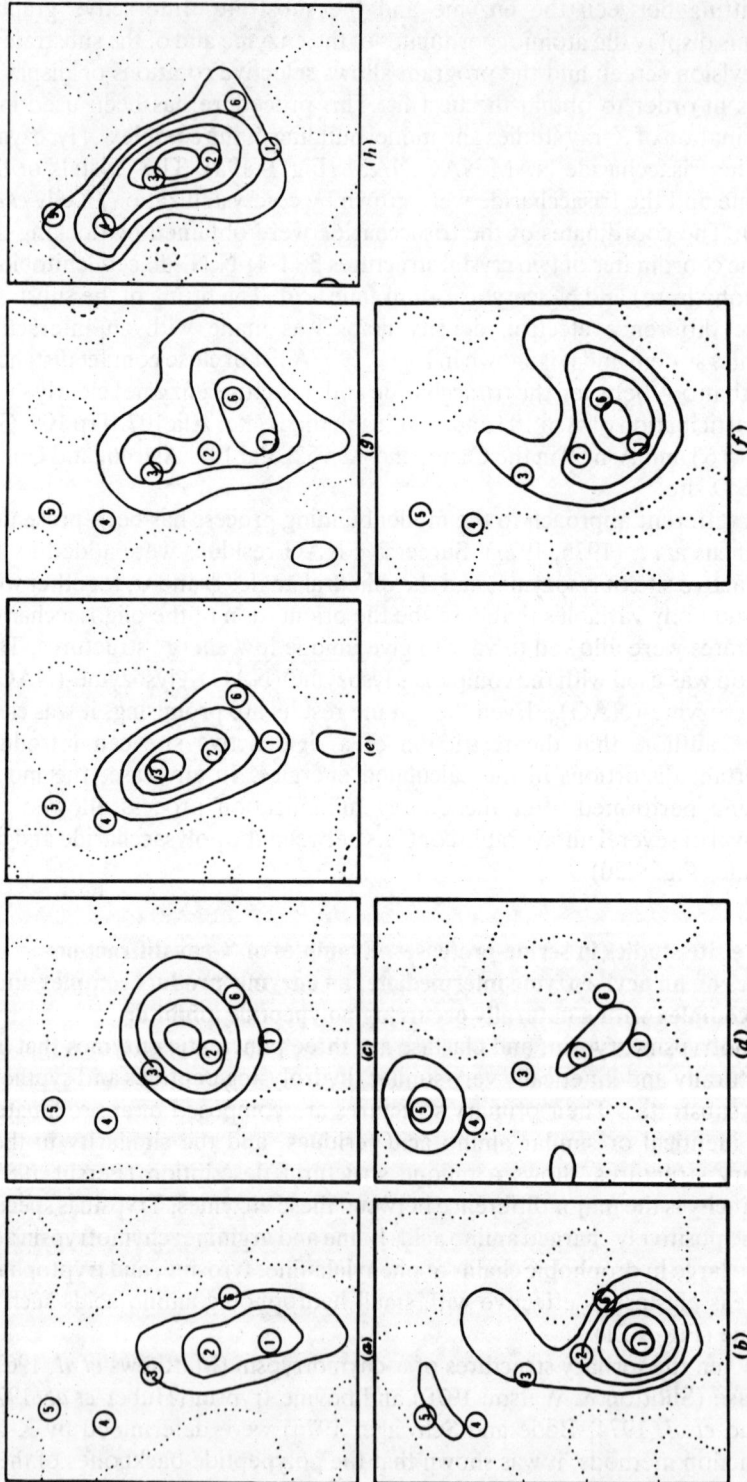

Fig. 9.17 — The section $y = 39/120$ in difference electron density maps at 6 Å resolution calculated with several lysozyme-inhibitor complexes and with the native enzyme. These show the binding of (a) NAG; (b) NAM; (c) 6-iodo-α-methyl -NAG; (d) α-benzyl-NAM; (e) and (g) di-NAG; (f) NAG-NAM; (h) (NAG)₃ to the enzyme. Apparent binding sites of amino-sugar residues are marked 1 to 6 (2 Å resolution studies showed later that sites 4, 3 and 2 correspond to binding sites A, B and C, respectively). (From Blake *et al.* 1967.)

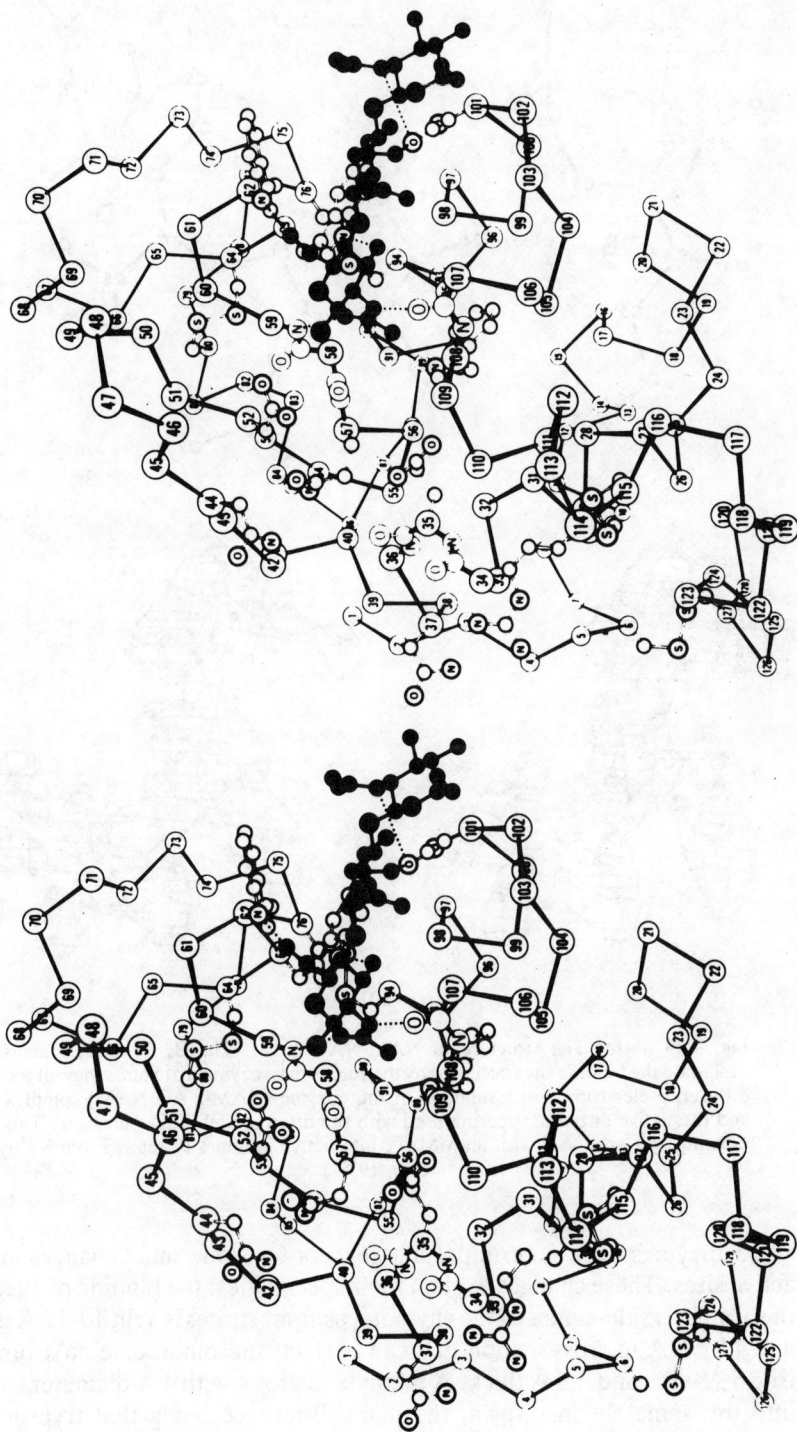

Fig. 9.18 — Stereo view of the α-carbon representation of the tertiary structure of lysozyme with the inhibitor (NAG)₃ positioned at the active site. (From Dickerson & Geis 1969.)

(a)

(b)

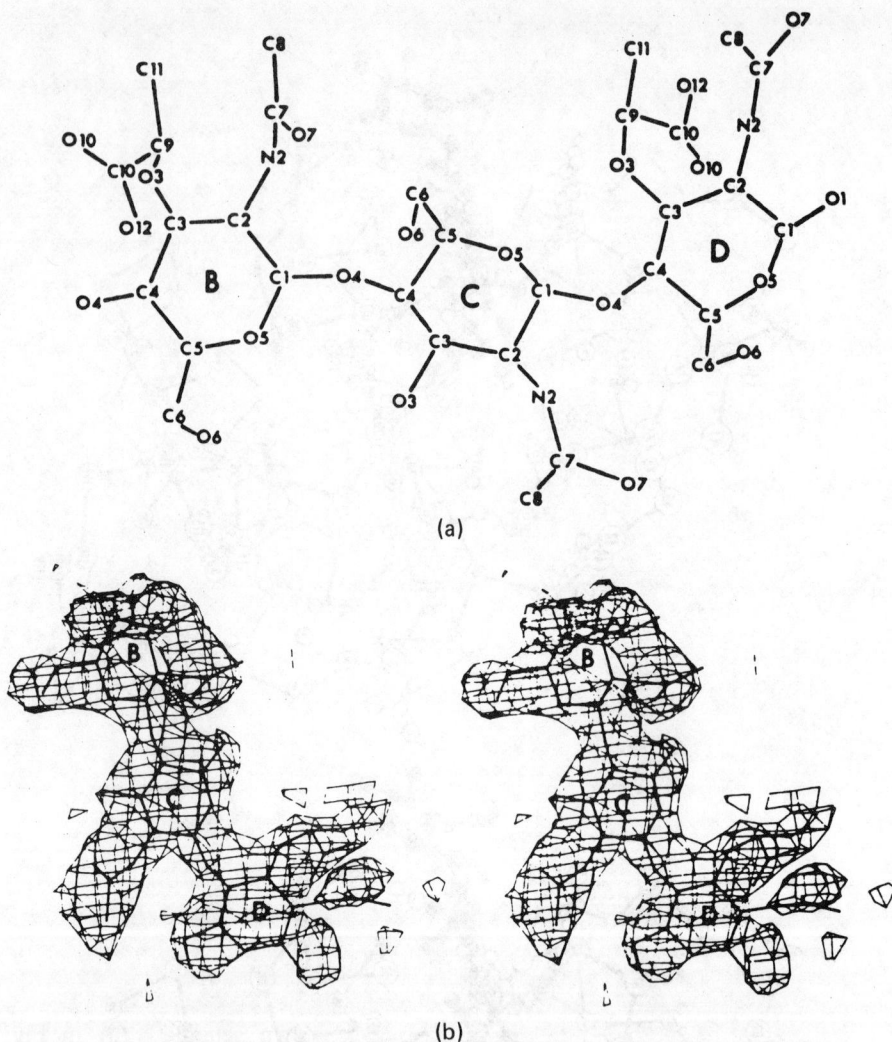

Fig. 9.19 — (a) The structure of NAM-NAG-NAM. The B, C, and D labels designate the binding sites occupied by the sugars in lysozyme. (b) Stereo view of the difference electron density function of the enzyme-(NAM-NAG-NAM) complex and the native enzyme, superimposed with the fitted model of the substrate. This drawing was produced with an MMS-X interactive graphics system. (From Kelly *et al*. 1979.)

three enzymes were superimposable, except for some small changes in the active sites. These changes explain their specificities: the binding pocket for the aromatic side-chains of the chymotrypsin substrates is a slit 10–12 Å deep and 4 by 6 Å in cross section that can perfectly accommodate an aromatic ring 6 Å wide and 3.5 Å thick. A methylene group with 4 Å diameter can fit into the same slit in trypsin, the main difference being that trypsin has replaced Ser 189 by Asp 189 to bind the positively charged side-chains of

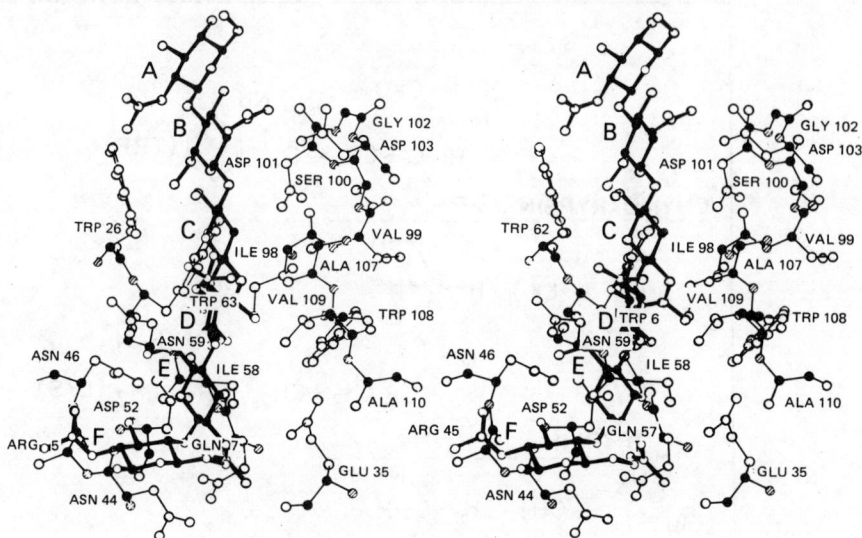

Fig. 9.20 — Stereo view of the calculated minimum energy structure of (NAG)$_6$ bound to the rigid active site of lysozyme. The substrate bonds are indicated in heavy lines. (From Pincus *et al.* 1977.)

lysine or arginine. The main changes in the active site of elastase with respect to chymotrypsin are the substitution of Gly 216 and Gly 226 by the bulkier Val 216 and Thr 226 groups. These changes make the active site accessible only to the small hydrophobic amino acid side-chain of alanine (Fig. 9.22).

The three enzymes belong to the group of the 'serine proteases', since it has been shown that a specific Ser residue (Ser 195) takes part in the catalytic process of each of these enzymes. The tertiary structures show that the imidazole ring of His 57 is hydrogen bonded to the carboxylate of Asp 102, to form a catalytic triad common to all serine proteases. The mechanism of hydrolysis for all serine proteases is shown schematically in Fig. 9.23 (Fastrez & Fersht 1973). A non-covalent intermediate is first formed, followed by attack of the hydroxyl group of Ser 195 on the substrate to give a tetrahedral intermediate. This intermediate collapses to give an acyl–enzyme, releasing the amino or alcohol. The acyl–enzyme then hydrolyzes to form the enzyme–product complex.

The tertiary structure of both the chymotrypsin product complex and the acyl–enzyme intermediate have been solved by X-ray diffraction methods. The enzyme–product complex was formed by diffusing into α-chymotrypsin crystals the acylated amino acids formyl-L-tryptophan and formyl-L-phenyl-lalanine. The interpretation of the difference Fourier maps at 2.5 Å resolution allowed the location of the acylated amino acid in the active site, as shown for formyl-L-tryptophan in Fig. 9.22. (Steitz *et al.* 1969).

The lifetime of an acyl–enzyme intermediate in the serine-proteases is of the order of 0.01 seconds at neutral pH, too short for crystallographic

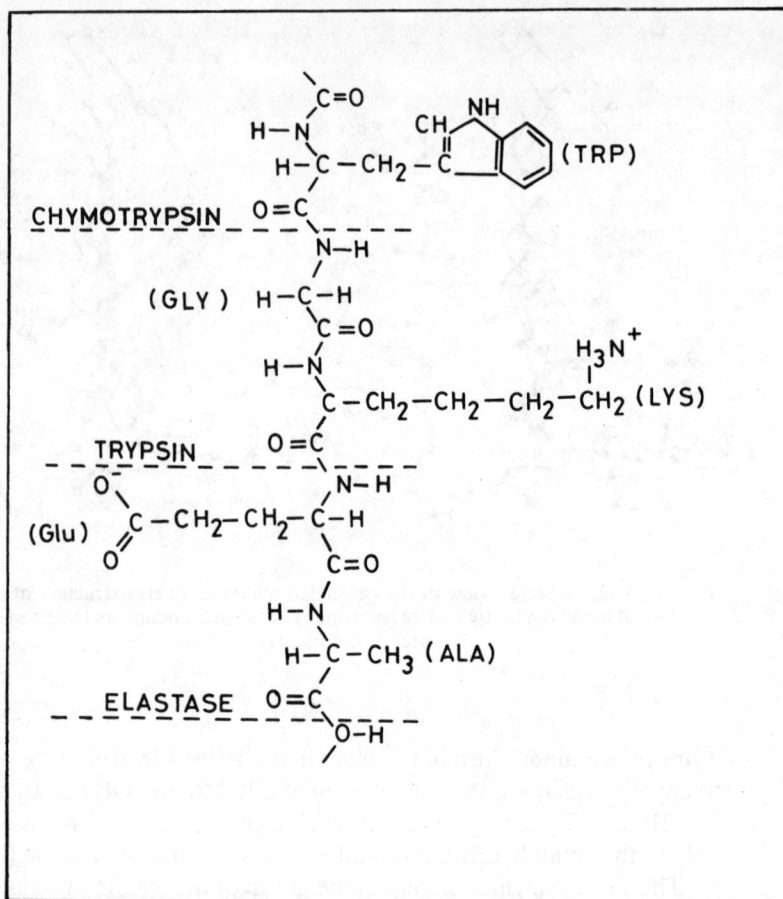

Fig. 9.21 — Sites of substrate cleavage for three serine proteases.

observations even with the most modern equipment. This limitation has been solved by using an abnormally poor substrate and a non-physiological pH to produce the acyl–enzyme indoleacryloyl-α-chymotrypsin. The crystals of the complex were obtained by incubation of α-chymotrypsin crystals in a saturated solution of indoleacryloyl-imidazole in 3% dioxane and 65% ammonium sulfate at pH 4.0 (Henderson 1970). The results obtained were extrapolated to the structure of the enzyme complex at neutral pH and lead to the proposal of an acylation mechanism for chymotrypsin.

There are many naturally occurring polypeptide inhibitors which bind to chymotrypsin and to trypsin producing stable enzyme–substrate complexes. They bind as true substrates, but their own structure does not allow the flexibility that a normal substrate shows on binding. These inhibitors lock the active site and do not allow the diffusion of the amino group released on the cleavage of the peptide from the enzyme. One of these inhibitors is the pancreatic trypsin inhibitor (PTI), a polypeptide with 58 amino acid residues

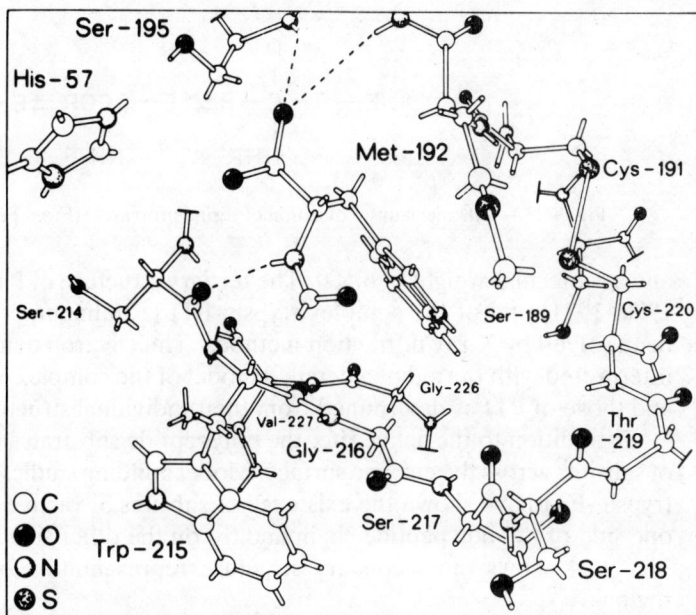

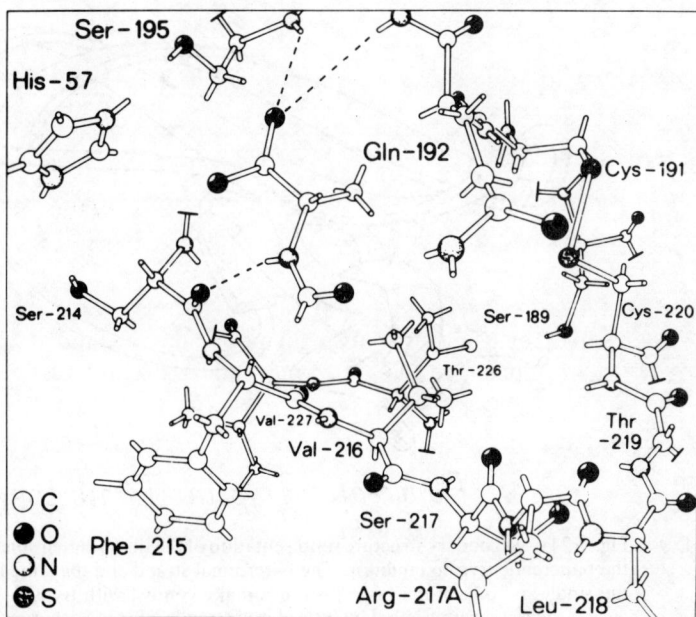

Fig. 9.22 — The active sites of chymotrypsin and elastase. The trypsin active site resembles that of chymotrypsin, except that Ser 189 becomes Asp 189. This difference allows trypsin to bind the positively charged Lys or Arg residues. (From Fersht 1985.)

$$E-OH+RCONHR' \rightleftharpoons E-OH.RCONHR' \rightleftharpoons$$

$$E-O-\overset{\overset{\displaystyle O^-}{|}}{\underset{\underset{\displaystyle NHR'}{|}}{C}}-R \rightleftharpoons \underset{+}{E-OCOR} \rightleftharpoons \underset{\underset{\displaystyle E-OH+RCO_2H}{\big\Updownarrow}}{E-OH.RCO_2H}$$
$$NH_2R'$$

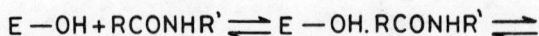

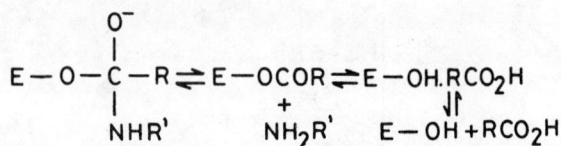

Fig. 9.23 — The mechanism of action of serine proteases. (From Fersht 1985.)

and a molecular weight of 6,500. The tertiary structure of PTI (Huber *et al.* 1970, 1971) and of the complex trypsin-PTI (Ruhlmann *et al.* 1973) have been solved by X-ray diffraction methods. The electron density maps were interpreted with the aid of a detailed model of the complex α-chymotrypsin and those of PTI as determined from their individual structures.

In addition to the active site, the polypeptide substrates bind to a series of subsites across the enzyme surface. Model building studies of the complex trypsin-PTI have shown the existence of subsites S_1 (active site), S_2, S_3 on one side of the polypeptide chain and S_1' on the other side (Fersht, 1985). Fig. 9.24 shows the secondary structure representation of PTI and the regions
where this inhibitor binds to trypsin.

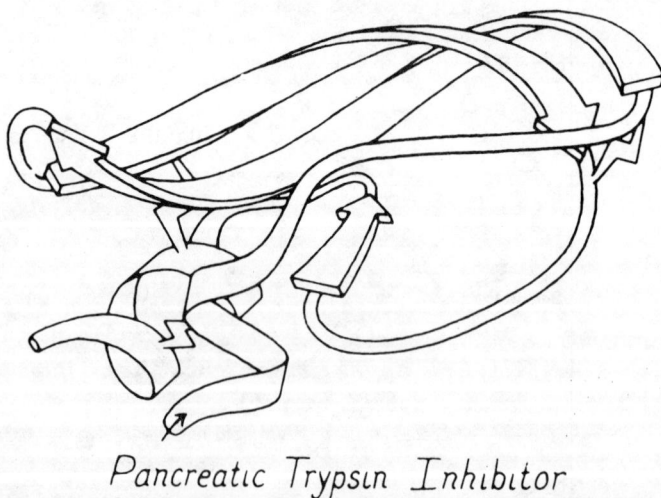

Pancreatic Trypsin Inhibitor

Fig. 9.24 — Secondary structure representation of the three-dimensional structure of the pancreatic trypsin inhibitor. The N-terminal strand and the large loop between the small and one of the large β-strands make contact with trypsin. The disulfide bridges are represented by little zigzag strands. (From Richardson 1985.)

CONCLUSIONS

At the beginning of this chapter we were faced with two problems: to observe a protein molecule with a resolution power of 10^{-7} mm, a million times larger than that of our naked eye, and, the necessity to 'freeze' an enzymatic reaction in order to discern all the intermediate complexes formed.

The solution to the first problem is given by the X-ray diffraction methods. The aim of this chapter is to give to non-crystallographers the basis of these methods together with some selected examples of their application to active site studies.

The examples were selected to illustrate different solutions to the problem of 'freezing' the enzymatic reaction, but these examples cover by no means all possible solutions to the study of enzyme–intermediate complexes.

This chapter also intends to show the power of the X-ray diffraction methods. The possibility of knowing the exact location of every atom in a molecule has drastically changed the trends of the biological sciences since the first tertiary structure of a protein was determined.

The X-ray diffraction methods also have some limitations. These limitations explain why it is not always possible to use these methods:

(1) It is a difficult technique which may require long years of patient and hard work to obtain and measure hundreds of diffraction spectra with thousands of diffraction spots each.
(2) The crystallization of a protein, the first step in the use of this method, is not a straightforward task even today. There are proteins that fail to form stable enzyme–heavy atom complexes, thus making the phase determination, necessary to calculate the electron density function, impossible.

These limitations have led to the development of a host of alternative methods for the analysis of enzyme active sites. Some of them are discussed elsewhere in this book.

REFERENCES

Artymiuk, P.J. & Blake, C.C.F. (1981) *J. Mol. Biol.* **52** 737–762.
Banner, D.W., Bloomer, A.C., Petsko, G.A., Phillips, D.C. & Pogson, C.I. (1971) *Cold Spring Harbor Symp. Quant. Biol.* **36** 151–155.
Banner, D.W., Bloomer, A.C., Petsko, G.A., Phillips, D.C., Pogson, C.I., Wilson, I.A., Corran, P.H., Furth, A.J., Milman, J.D., Offord, R.E., Priddle, J.D. & Waley, S.G. (1975) *Nature (London)* **255** 608–614.
Bergstén, P.C., Waara, I., Lövgren, S., Liljas, A., Kannan, K.K. & Bengtsson, U. (1972) *Proc. of the Alfred Benzon Symp. IV* (Rørth, M. and Astrup, P., eds) Munksgaard, Copenhagen, 363–383.
Blake, C.C.F., Koenig, D.F., Mair, G.A., North, A.C.T., Phillips, D.C. & Sarma, V.R. (1965) *Nature (London)* **206** 757–761.
Blake, C.C.F., Johnson, L.N., Mair, G.A., North, A.C.T., Phillips, D.C. & Sarma, V.R. (1967) *Proc. Royal Soc. London* **B167** 378–388.
Bode, W. & Schwager, P. (1975) *J. Mol. Biol.* **98** 693–717.
Fastrez, J. & Fersht, A.R. (1973) *Biochemistry* **12** 2025–2041.
Henderson, R. (1970) *J. Mol. Biol.* **54** 341–354.

Huber, R., Kukla, D., Rühlman, A., Epp, O. & Formansk, H. (1970) *Naturwiss.* **57** 389–392.

Huber, R., Kukla, D., Rühlman, A. & Steigemann, W. (1971) *Cold Spring Harbor Symp. Quant. Biol.* **36** 141–150.

Huber, R., Kukla, D., Bode, W., Schwager, P., Bartels, K., Deisenhofer, J. & Steigemann, W. (1974) *J. Mol. Biol.* **89** 73–101.

Kannan, K.K., Nostrand, B., Fridborg, K., Lövgren, S., Ohlsson, A. & Petef, M. (1975) *Proc. Natl. Acad. Sci. USA* **72** 51–55.

Kannan, K.K., Petef, M., Fridborg, K., Cid-Dresdner, H. & Lövgren, S. (1977) *FEBS Letters* **73** 115–119.

Kelly, J.A., Sielecki, A.R., Sykes, B.D., James, M.N.G. & Phillips, D.C. (1979) *Nature (London)* **282** 875–878.

Knowles, J.R., Leadley, P.F. & Meister, S.G. (1972) *Cold Spring Harbor Symp. Quant. Biol.* **36** 157–164.

Lindskog, S., Henderson, L.E., Kannan, K.K., Liljas, A., Nyman, P.O. & Strandberg, B. (1971) *The Enzymes (3rd Edition)* **5** 587–665.

Lövgren, S., Bergstén, P.C., Liljas, A., Petef, M. & Bengtsson, U. (1971) cited by Bergstén *et al.* 1972.

Matthews, B.W., Sigler, P.B., Henderson, R. & Blow, D.M. (1967) *Nature (London)* **214** 652–656.

Nostrand, B. (1974) PhD thesis, University of Uppsala, Sweden.

Nostrand, B., Waara, I. & Kannan, K.K. (1975) *Isozymes I. Molecular Structure*, pp. 575–599, Academic Press, New York .

Phillips, D.C. (1967) *Proc. Natl. Acad. Sci. USA* **57** 484–495.

Pincus, M.R. Zimmerman, S.S. & Scheraga, H.A. (1976) *Proc. Natl. Acad. Sci. USA* **73** 4261–4265.

Pincus, M.R., Zimmermann, S.S. & Scheraga, H.A. (1977) *Proc. Natl. Acad. Sci. USA* **74** 2629–2633.

Richardson, J. (1985) *Methods Enzymol.* **115** 341–358.

Rose, I.A. (1975) *Advances Enzymol.* **43** 491–517.

Ruhlmann, A., Kukla, D., Schwager, P., Bartels, K. & Huber, R. (1973) *J. Mol. Biol.* **77** 417–436.

Shotton, D.M. & Watson, H.C. (1970) *Nature (London)* **225** 811–816.

Steitz, T.A., Henderson, R. & Blow, D.M. (1969) *J. Mol. Biol.* **46** 337–348.

Stroud, R.M., Kay, L.M. & Dickerson, R.E. (1974) *J. Mol. Biol.* **83** 185–208.

BOOKS RECOMMENDED FOR FURTHER READING

Blundell, T.L. & Johnson, L.N. (1976) *Protein Crystallography*, Academic Press, New York.

Buerger, M.J. (1960) *Crystal Structure Analysis*, John Wiley and Sons, New York.

Dickerson, R.E. & Geis, I. (1969) *The Structure and Action of Proteins*, Harper & Row, New York.

Fersht, A. (1985) *Enzyme Structure and Mechanism*, 2nd edn, W.H. Freeman, San Francisco.

Harburn, C., Taylor, C.A. & Welbery, T.R. (1975) *Atlas of Optical Transforms*, G. Bell and Sons, London.
Lipson, H. (1972) *Optical Transforms*, Academic Press, London.
Ramachandran, G.N. & Srinivasan, R. (1970), *Fourier Methods in Crystallography*, Wiley-Interscience, New York.

10

Applications of nuclear magnetic resonance to the study of enzyme active-sites

Dr Thomas Nowak, Department of Chemistry, University of Notre Dame, Notre Dame, IN 46556, USA

INTRODUCTION

Nuclear magnetic resonance (NMR) spectroscopy can, in principle, yield detailed information regarding enzyme structure and the structure of the specific ligands which bind to the enzyme. The structure of the ligands at the binding site of enzymes and the structure of enzyme–ligand complexes can also be obtained. Since the NMR phenomenon is a time dependent phenomenon, kinetic as well as thermodynamic and structural information regarding both the enzymes and the substrates can be obtained. The attraction of NMR is that (again, in principle) one can investigate the magnetic nuclei of each of the atoms within the molecule of the enzyme (^1H, ^{13}C, ^{14}N...), of the ligands which bind to the enzyme (^1H, ^{19}F, ^{31}P, ^{13}C...), or of the environment of the active-site (solvent H_2O, ^{23}Na, ^{39}K, ^{35}Cl...). Since a large number of enzymes either contain metal ions (metalloenzymes) or require the addition of metal ions for activity (metal-requiring enzymes), a variety of these metal ions can be observed by NMR. These metals include divalent cations (^{25}Mg, ^{43}Ca, ^{59}Co, ^{113}Cd...) and monovalent cations (^7Li, ^{23}Na, ^{39}K, ^{205}Tl...). The rapid advance in technology of high-resolution, high-field super-conducting magnets, radio frequency electronics, and computer systems have given birth to several new generations of high-field, high-resolution, multinuclear NMR spectrometers. Such instruments can allow experiments and measurements to be made which were nearly unimaginable only a decade ago.

A further attraction of the NMR technique is that ligand binding, structural changes, and environmental changes at the catalytic site of an enzyme can be studied regardless of whether the enzyme is catalytically active, partially active, or inactive. This ability, currently highly underused, can yield important information concerning the function of specific amino acids in ligand (substrate, metal activator, hetrotropic modulator...) binding and in the catalytic processes. This chapter will attempt to discuss the utilization of NMR spectroscopy in conjunction with other methods—primarily chemical modification—to determine active-site structure. Because of the relative ease in studying metal ion effects, and in the utilization of certain metal ions as probes in the study of enzyme and enzyme–ligand interactions, these applications may appear to be inordinately stressed in this chapter. The basic approaches will be examined, and the limitations discussed. Some general familiarity with the NMR method will be assumed, and general references will be given but, they will not be exhaustive. Theory will be discussed only as deemed necessary to comprehend and to utilize the method in a practical manner. More detailed background can be obtained from the references cited.

NMR yields three general parameters which are useful in obtaining information regarding the structure and dynamics of the system under investigation. The chemical shift (δ) of a resonance which is observed is a function of the magnetic environment of the nuclei being investigated. This property makes NMR spectroscopy a potent tool in the study of molecules and their structure. If all protons within a molecule absorbed radio frequency power at exactly the same frequency, NMR would have little utility. The phenomenon of a chemical shift arises owing to shielding of the nuclei under examination from the applied magnetic field by the electrons. Thus it is the electronic environment that causes variations in chemical shift. Any factor that will alter the electron density at the nucleus will alter the chemical shift. Shielding of methyl protons is greater than that of methylene protons, and still greater than that of aromatic protons, for example. Thus the resonance of a methylene proton is further upfield than that of a proton on an aromatic system, and a methyl proton is furthest upfield. If spectra are obtained on samples that are fully relaxed and additional effects such as Overhauser effects do not occur, the area under the peak for each resonance is directly proportional to the concentration of nuclei. Both the relative and, in some cases, absolute distribution of magnetically non-equivalent nuclei and contaminant levels can be quantitated.

The second parameter is the spin–spin coupling or scalar coupling constant, $J_{i,j}$, that occurs between two nuclei of spin I, I_i and I_j. The term I is the nuclear spin or spin angular momentum which has integer and half integer values $(I = 0, \frac{1}{2}, 1, \frac{3}{2}...)$. Spin I_i is split into $2nI_j + 1$ lines by spin j. Spin j is split into $2nI_i + 1$ lines by spin i. The term n indicates the number of nuclear spins. Thus, if spin i is one ^{13}C $(I = \frac{1}{2})$ and spin j is two ^{1}H's $(I = \frac{1}{2})$ in a methylene group, then the ^{13}C spectrum has three lines and the ^{1}H spectrum has two. The magnitude of the coupling constant, $J_{i,j}$, depends upon the interactions between the spins. Since most coupling is observed with nuclei of spin $= \frac{1}{2}$, discussion is primarily limited to such cases. Coupling constants depend on the environment of the molecule and the relative orientation or molecular geometry of the nuclei under observation, and therefore are important in structure determination. These coupling constants are independent of the magnetic field. The closer the nuclei are to each other (fewer bonds), the larger the magnitude of the coupling for related molecules. There are certainly cases, however, where three-bond coupling constants are larger than two-bond coupling constants. If the chemical shifts or effective chemical shifts of the coupled nuclei are large compared to the coupling constant, then the spectral patterns are relatively simple and are considered first-order. When the chemical shifts are of the magnitude of the coupling constant, the spectra become more complex and are called second-order. Resolution of coupling is an important spectroscopic technique in structure determination. Spin–spin coupling can be studied by double resonance, spin–decoupling experiments, spectral simulation and by two dimensional correlation spectroscopy (Becker 1980).

The third and most often neglected of the parameters are the relaxation rates of the nuclei. In fact, in the initial search for a nuclear resonance phenomenon, dynamic processes and line shapes were of primary interest, and coupling constants and chemical shifts observed in liquids came as a surprise. The equations derived to define the motion of the magnetic moment (μ) or magnetization M in the samples were given by Bloch (1946). The motion in the direction of the external magnetic field H_o is designated as $dM,z/dt$. In the plane perpendicular to H_o, the x,y plane, the motion of the magnetization vector is designated as $dM,x/dt$ and $dM,y/dt$. Magnetization in the x,y plane occurs because of the property of spin of the nuclei. When a sample with a nuclear spin is unaffected by a magnetic field, the magnetic moment is zero. When a nuclear spin is placed in an external magnetic field, H_o, a torque is placed on the magnetic moment M by H_o to change the angular momentum, P

$$\frac{dP}{dt} = -H_o x M \tag{1}$$

Since the spin angular momentum is related to the magnetic moment by the magnetogyric ratio τ,

$$M = \gamma P \tag{2}$$

then

$$\frac{dm}{dt} = -\gamma H_o M \tag{3}$$

This expression describes the motion of the magnetic moment or magnetization about the z axis defined as the direction of the H_o field. At equilibrium the nucleus has a magnetization of M_o. The decay or relaxation of the magnetization in the z axis is characterized by a relaxation rate, $1/T_1$. A change in M_z is accompanied by a transfer of energy between the nuclear spin and other degrees of freedom or the lattice of the surroundings and is hence called the 'longitudinal relaxation rate' or the 'spin–lattice relaxation rate', $1/T_1$. A decay in the transverse components of the magnetization, M_x and M_y, results in an exchange of energy between spins of different nuclei without transfer to the lattice, and is called the 'transverse relaxation rate' or the 'spin–spin relaxation rate', $1/T_2$. In solution studies, both T_1 and T_2 are affected by the exchange of energy between the spin system being studied and the environment. Since energy exchange between spin systems is dependent upon dipolar effects, distances between these dipoles can be calculated if the magnitudes of these effects can be measured. Since these relaxation phenomena are time-dependent, kinetic information such as molecular motion is possible from these studies. More detailed treatments are available (Abragam 1973, James 1975).

ENZYME STUDIES

In the study of enzymes it is conceivable that a ^1H spectrum of the enzyme can yield absorption peaks for each of the protons in the molecule. In order to do such an experiment, several problems must first be considered. A proton-free solvent and proton-free buffer are, in general, desired. The solvent of choice which should give a minimum of perturbation of the protein structure is deuterium oxide, ^2H_2O or simply D_2O. To attempt protein NMR studies, there are a variety of NMR methods to suppress solvent signals (suppress the signal from H_2O or DHO), each with their inherent advantages and disadvantages (Turner 1984). The D_2O commercially available can be obtained with 0.1% protons or less. If a 99.99 atom % D_2O solvent is used, the proton content of the solvent is 11 mM, higher than the possible concentration of most enzymes.

Little information is available concerning the effects of D_2O on protein structure; however, it is common that a solvent isotope effect on enzymatic activity is observed. Reasons for such effects are often difficult to assess (Schowen 1972). Buffer systems can also be a problem. If a phosphate buffer causes no problem with the enzyme under investigation, it will be ^1H NMR invisible. Otherwise either low concentrations of buffer or no buffer may be used.

The two major problems with this NMR approach are the concentration of enzyme and resolution of the spectra. The signal-to-noise of the spectrum is directly proportional to the concentration of the sample. Many enzymes may not be sufficiently soluble to yield a 1×10^{-3} M solution. If 5 mm sample tubes are used, 0.30 ml is a minimum volume to obtain spectra. If a protein of molecular weight 1×10^5 is used, this requires 30 mg of enzyme for a spectrum. Even if solubility is not a major problem, an increase in concentration increases the viscosity of the sample. In more viscous solutions rapid averaging of the sample no longer occurs and broad absorption lines are observed, which decreases resolution of the spectrum.

In an enzyme of molecular weight approximately 70 000 (an 'average' size protein) the rotational correlation time, τ_r in aqueous solution may be estimated at 10^{-8} s using the Stokes–Einstein equation, assuming the protein is roughly globular. This enzyme is also expected to contain approximately 600 amino acids. The large number of residues results in a high number of overlapping resonances because of the number of protons present. A 'typical' ^{1}H spectrum shows a large envelope of overlapping resonance peaks upfield from the water resonance, that results from the aliphatic groups in the protein. An envelope downfield from the water resonance is due to the aromatic groups on the protein. The resonances that appear are usually quite broad as well. The broad lines are caused by dipole–dipole interactions with either the same or with other nuclei. The effect of dipolar interactions on line widths ($1/\pi T_2$) is modulated by the rotational correlation time of the group under investigation. An increase in the line widths of resonances in small molecules with decreasing tempera-ture is often seen for the same reason. Regardless of the nature of the dipoles that affect the relaxation, $1/T_2$ is directly proportional to τ_r. If some groups have less restrictive motion in the enzyme they may yield sharper peaks that may protrude from the envelope of broad, overlapping peaks. Although assignments of resonances of free amino acids have been made (Roberts & Jardetzky 1970), the assignments of resonances which may be observed for an enzyme must be made for specific amino acid residues within the enzyme structure. This can be a severe limitation. Often an approach such as specific amino acid derivatization prior to obtaining the spectrum can help in making assignments. New multi-pulse methods can aid in structure determination, but to date there is little such detailed information on molecules as complex as proteins (Wüthrich *et al*. 1982).

The most useful approach to studying enzyme structure by protein NMR with a minimum of perturbation has been the observation of the resonances from histidine. The C-2 and C-5 proton resonances are downfield from the aromatic protons. If a limited number of these amino acids are present in the enzyme, they have sufficient molecular motion (to yield 'sharp' lines), and are in different (magnetic) environments in the enzyme, they can yield reasonably sharp, resolved resonances. The classical use of these properties was with the small enzyme RNAase (M_r = 24 500) (Meadow & Jardetzky 1968), and recently the large enzyme (M_r = 237 000) pyruvate kinase was so studied (Meshitsuka *et al*. 1981). The C-2 proton resonance is especially sensitive to the ionization state of the imidazole nitrogens, thus the pK_a for each individual histidine within the native enzyme can be obtained from titration studies. It remains that proper assignments be made for each histidine. The binding of a ligand or metal ion to a specific histidine or histidines could result in a change in the magnetic environment (chemical shift) of the resonance and an alteration in the pK_a. This application of NMR has usefulness in some limited number of enzymes.

It is possible that ^{13}C and ^{14}N studies can also be performed. An increase in the range of chemical shifts for these nuclei enhances spectral dispersion and increases the possibility of resolving more resonances. A major problem with these nuclei is the low natural abundance (1.1%) and sensitivity (1.6%) for ^{13}C and the low sensitivity (0.1%) for ^{14}N compared to the sensitivity of ^1H (100%). The quadrupolar ^{14}N nucleus also has a nuclear spin of 1 which gives substantial line broadening. With enzymes from bacterial systems where it is feasible to consider growing the organism on media or precursors (i.e. amino acids) that are selectively enriched (^{13}C or ^{15}N) (Hunkapiller et al. 1973), some of these studies may become more reasonable. New DNA cloning techniques can expand this potential. A recent detailed review of ^{13}C NMR studies of enzymes has been published (Malthouse 1986).

An alternative approach to looking at the enzyme in an effort to obtain information regarding enzyme structure and the effects of ligand binding on the enzyme is to use a reporter group on the enzyme. One of the more sensitive groups that can be studied is ^{19}F. The use of this nucleus in enzyme systems has been reviewed (Gerig 1981). This nucleus is 83% as sensitive as ^1H, has a large range of chemical shifts, is rather sensitive to its magnetic environment, and there are no background resonances of ^{19}F which cause interference. A ^{19}F reporter group can be incorporated by one of two methods. A specifically fluorinated amino acid (i.e. fluorotyrosine, fluoro-alanine) can be added to growth medium and incorporated into the protein (Sykes & Weiner 1980). Under these conditions one group of amino acids (i.e. tyrosines, alanines) would contain the ^{19}F resonance. Most organisms will not grow on 100% fluorinated amino acids as they are toxic at higher levels. Furthermore, each of the residues is labeled and will exhibit a resonance. In a case where each residue is non-equivalent, assignments for each residue (i.e. each tyrosine) may be necessary.

An alternative to this approach is to covalently label the enzyme at a specific residue with a fluorine-containing reagent. Among the possible reagents one may use are trifluoroacetic anhydride, trifluoroacetyliodide, or 3-bromo-1,1,1-trifluoropropanone. The chemical shift and/or the line width ($1/T_2$) of the ^{19}F label, a 'reporter' for a change in the enzyme structure, must reflect ligand binding and/or catalysis. If the ^{19}F resonance is sensitive to conformational changes in the enzyme then site-specific modification of groups at the active-site will be reflected by changes in the ^{19}F resonance. Ligand binding to modified enzyme may also be monitored by a measure of the spectral parameter (δ or $1/T_2$) as a function of ligand concentration. A titration of the spectral parameter versus ligand concentration yields a titration curve that is evidence for ligand binding. A dissociation constant for ligand binding can be determined.

The method of using reporter groups can be expanded with other labels. Most other labels would be less sensitive than fluorine; however, the modification may be more selective or may yield reporter groups that are more sensitive to changes in enzyme structure. [^2H] labels or [^{13}C] labels can also be incorporated into the protein. A potential strength of using these labels is that the incorporation of ^2H for ^1H or ^{13}C for ^{12}C into the protein will have a very minor, if any, effect on the protein itself. Although reporter groups yield information regarding the environment of the group and not specific structural features of the enzyme, comparative structural changes can be studied by such methods.

The method of photo-chemically induced nuclear polarization (photo-CIDNP) which originates from free radical reactions has recently been developed as a sensitive method to measure structural changes on the surface of proteins (Kaptein 1982). The method requires a modified spectrometer and a proper light source (laser) to begin to probe surface changes. These changes, when observed, are reflected in changes about aromatic amino acids. This technique has the advantage of high sensitivity, and it yields general conformation information.

STUDIES OF THE LIGANDS

An alternative to measuring aspects of the enzyme and its structure in the study of enzyme–ligand interactions is an investigation of the ligand itself. A general definition of a ligand implies substrates, modifiers, inhibitors, and activators including metal ions. The proper studies depend upon the enzyme of interest. There are two potential types of experiment one can perform. In some cases the interaction of a ligand with an enzyme results in the formation of an enzyme–ligand complex such that partial immobilization of a portion of the ligand occurs. A decrease in the mobility of a group (i.e. a methyl group) increases the correlation time, the time constant for the process that modulates or interferes with the relaxation process. The rotational correlation time τ_r of the methyl group is the rotation time of that group which modulates the dipolar interactions among the methyl protons and results in an increase in $1/T_2$ and $1/T_1$. The $1/T_2$, estimated from the line width of the resonance, is the parameter that is more easily measured. If the effect on $1/T_2$ is sufficiently large and the ligand is in the fast exchange domain (the lifetime, τ_m, of the ligand in the E–L complex is short compared to the relaxation time of the nucleus, $T_{2,b}$, in the E–L complex) an average line width ($1/T_{2,obs}$) for the bound ligand ($1/T_{2,b}$) and free ligand ($1/T_{2,0}$) is observed.

$$\frac{1}{T_{2,obs}} = \frac{[L_b]}{[L_T]}\left(\frac{1}{T_{2,b}}\right) + \frac{[L_f]}{[L_T]}\left(\frac{1}{T_{2,0}}\right) \tag{4}$$

The observed effect is a mole average effect of bound ligand $[L_b]$ and free ligand $[L_f]$ where the sum of the concentrations of L_b and L_f give the concentration of total ligand, $[L_T]$. From a determination of the amount of ligand bound (the concentration of enzyme sites if the enzyme is saturated with ligand) and the total amount of ligand present, $1/T_{2,b}$ can be calculated. Values for $1/T_1$ can be handled by similar treatment if $1/T_{1,obs}$ is measured. If the dipolar effect is all intramolecular and the nature of the dipoles is known (e.g. ^1H-^1H interactions), the value for the rotational correlation time for that group in the enzyme–ligand complex can be calculated. From a determination of ligand binding, values for [L_b] and [L_f] can be obtained and $1/T_{1,b}$ and $1/T_{2,b}$ calculated. From the structure of the molecule, the distance r between the dipoles is usually obtained. The distance r is estimated from crystal structure data or from models of such compounds.

$$\frac{1}{T_{1,b}} = \frac{3\gamma_I^4\hbar^2}{10r^6} \left(\frac{\tau_c}{1 + \omega_I^2\tau_c^2} + \frac{4\tau_c}{1 + 4\omega_I^2\tau_c^2} \right) \tag{5}$$

$$\frac{1}{T_{2,b}} = \frac{3\gamma_I^4\hbar^2}{20r^6} \left(3\tau_c + \frac{5\tau_c}{1 + \omega_I^2\tau_c^2} + \frac{2\tau_c}{1 + 4\omega_I^2\tau_c^2} \right) \tag{6}$$

In these equations, $\hbar$ is Planck's constant/π, and τ_c, the correlation time for the dipolar interactions, is τ_r. The value ω_I is the Larmor frequency in rad sec^{-1}. If such immobilization is detected and calculated for the ligand bound to the native enzyme, then one can determine if immobilization of the same ligand occurs with modified enzyme. Restriction of molecular motion is one possible mechanism of catalytic activation.

Another approach to the study of ligand binding to enzymes is by the use of paramagnetic probes on the enzyme. The use of paramagnetic species to probe ligand interactions is feasible because an unpaired electron is about 657 times more effective than a proton in causing a dipolar effect on relaxation. Several approaches can be utilized to take advantage of these large dipolar effects. Stable nitroxides, many of which are commercially available (e.g. from Aldrich Chemical Co. and Merck & Co.) can potentially be covalently attached to the enzyme. These nitroxides include derivatives of iodoacetate, N-ethylmaleimide, and diisopropylfluorophosphate that can be used to label reactive groups such as cysteine, histidine, lysine, or reactive serine (Berliner 1976). Selectivity of labeling and choice of amino acid residue, discussed elsewhere in this text, is necessary. These probes can be monitored by EPR spectroscopy, or their effects on ligands can be studied by NMR. This label can be used as the reference point to study ligand interactions to labeled enzyme.

Alternative paramagnetic species that can be used are metal ions. These metals may either bind to the enzyme, are found enzyme–bound, or can bind as a metal–substrate complex to the enzyme. Some of the metal ions that can be utilized or substituted for the 'physiological' cation are Mn(II), Fe(II), Co(II), Cu(II), Gd(III), or Cr(III). If the enzyme being studied gives the investigator a choice of cations there are distinct advantages to using a few of these cations, particularly Mn(II), as will be shown. A determination of the stoichiometry of the paramagnetic center is necessary. With the nitroxide 'spin label' an integration of the EPR spectrum of labeled enzyme to obtain a spin count can be used. A comparison of the spectrum of the sample with a spectrum of a known spin label can be made. This is often the method of choice. In the case of metal ions the investigator has a variety of techniques available to measure concentration. With tight binding metals, atomic absorbtion spectroscopy can be used to determine the metal content of the enzyme for any metal ion. Alternatively, metal binding using unstable nuclei can be performed using one of a variety of equilibrium techniques such as equilibrium dialysis, gel permeation, ultrafiltration, ... The Mn(II) cation is almost uniquely suited for EPR studies where a solution spectrum of the free cation can be measured, and it yields a simple six-line spectrum. Upon ligand binding (ligand implying anything from a small molecule such as orthophosphate or ADP to protein) the change in zero field splitting and line broadening results in a 'disappearance' in the spectrum of bound Mn(II). The remaining signal is due to the free Mn(II) and the intensity of the spectrum is directly proportional to the concentration of free Mn(II) (Cohn & Townsend 1954). Proper binding studies will lead to a determination of the dissociation constant for the label and its stoichiometry per enzyme or enzyme active-site.

In most cases the metal ion utilized is either the physiologically important cation activator or can substitute for the physiologically important activator to elicit catalysis. The paramagnetic center is at the activator site which may be either at, near, or remote from the active-site. Other probes such as the lanthanides (e.g. Gd III) may serve as activators in a few cases or as inactive analogues that are competitive with the physiologically relevant cation. The lanthanide metals, in spite of the fact that they are most commonly trivalent, have f shell electrons which give nearly all of them interesting spectroscopic properties. For many NMR studies the physical properties of Gd(III) make it most useful. The Cr(III) cation which forms exchange inert ligand–metal complexes can also be used as a probe. This metal has recently found use as a kinetic and an NMR probe by being used as a Cr(III)-nucleotide complex (Cleland & Mildvan 1979). This metal nucleotide complex is an analog of Mg-nucleotide or Ca-nucleotide complexes that serve as substrates.

The paramagnetic probes, particularly nitroxides, Mn(II), Gd(III), and Cr(III), can have a substantial effect on the longitudinal and the transverse relaxation rates of the nuclei of the ligands that are in close proximity to the paramagnetic center. In the studies of enzyme active-sites by chemical modification, the use of such probes may be of exceptional value. After modification of the enzyme one can first determine if the binding site for the paramagnetic probe is still intact. Equilibrium binding or EPR binding (of Mn(II)) can determine if there is any alteration in the stoichiometry or in the dissociation constant for the cation to the modified enzyme. If the cation binding sites remain intact in the enzyme, then ligand binding to the modified enzyme can be studied. The results of a proper series of NMR experiments can describe the alteration in the binding of the ligands to the modified enzyme, the structure of the ligands at the binding site, and their exchange rates. This information can be compared to what is known regarding the structure and dynamics of ligand binding with the native enzyme to determine the effects of modification. Again, these studies can be performed even if the modified enzyme is totally inactive.

The effect of the paramagnetic species on the relaxation rates of the nucleus/nuclei in question must first be quantitated. The choice of nucleus studied is often dictated by the nature of the enzyme, its ligands, and ease of experimentation. For example, if the interaction of ATP to an enzyme–metal complex will be investigated, the ^{31}P nuclei of ATP are probably of most interest and are relatively easy to detect. The ^{1}H nuclei of the ribose portion of ATP yield a complex spectrum with overlapping lines and the resonances of the individual protons are much more difficult to resolve. The ^{13}C nuclei are of low natural abundance, and in an unenriched sample the experiments would take an inordinately long time.

To quantitate the paramagnetic effect of the probe on the relaxation rate of the nuclei, the relaxation rates are measured in the absence of the paramagnetic species $(1/T_{1,0}, 1/T_{2,0})$. This may be performed by a measurement of the nuclei in the presence of enzyme but no added metal, a diamagnetic metal (Mg(II), Zn(II), Ca(II)...), or with a reduced nitroxide label. The addition of the paramagnetic species is made by either adding the paramagnetic metal to the analytical sample that contains ligand and apoenzyme, or by adding the enzyme–metal complex to the solution. The procedure of choice depends upon the properties of the enzyme. If the enzyme is a metalloenzyme the latter approach can be used. If the enzyme is metal-requiring, then sufficient apoenzyme is present such that when metal is added most if not all of the metal binds to the enzyme. If a spin–label enzyme is added, since most spin labels are covalently attached to the enzyme, the labeled enzyme is added in increments. The observed relaxation rate $(1/T_{1,\mathrm{obs}}, 1/T_{2,\mathrm{obs}})$ is a function of the diamagnetic relaxation rate and the paramagnetic relaxation rate:

$$\frac{1}{T_{1,\mathrm{obs}}} = \frac{1}{T_{1\mathrm{p}}} + \frac{1}{T_{1,0}} \tag{7}$$

$$\frac{1}{T_{2,\mathrm{obs}}} = \frac{1}{T_{2\mathrm{p}}} + \frac{1}{T_{2,0}} \tag{8}$$

The paramagnetic effect is measured as a function of the concentration of paramagnetic species. If possible, a plot of $1/T_{i,\mathrm{obs}}$ vs the concentration of paramagnetic species can be made to show expected linearity in the relaxation rate where $i = 1$ or 2. The rate can be normalized for the concentration of the ligand, [L], and for the concentration of the paramagnetic species [p] by the term f where $f = [\mathrm{p}]/[\mathrm{L}]$. The normalized paramagnetic effects to the relaxation rates are related to the number of ligands (q) which bind to the specific site(s) in the vicinity of the paramagnetic probe, the relaxation time of the nucleus at this site (T_{iM}), and the lifetime of the nucleus of this site (τ_m). In some cases with paramagnetic ions a chemical shift change, Delta omega, is also observed which affects T_2 relaxation.

These effects have been described by Swift & Connick (1962) and by Luz & Meiboom (1964).

$$\frac{1}{fT_{1p}} = \frac{q}{T_{1M} + \tau_M} \tag{9}$$

$$\frac{1}{fT_{2p}} = \frac{q}{\tau_m} \left[\frac{\dfrac{1}{T_{2M}}\left(\dfrac{1}{T_{2m}} + \dfrac{1}{\tau_M}\right) + \Delta\omega^2}{\left(\dfrac{1}{T_{2M}} + \dfrac{1}{\tau_m}\right)^2 + \Delta\omega^2} \right] \tag{10}$$

If chemical shift changes are negligible or absent, eq. (10) reduces to

$$\frac{1}{fT_{2p}} = \frac{q}{T_{2M} + \tau_m} \tag{11}$$

In most such cases $\Delta\omega \approx$ zero, and eq. (11) can be used. The enzyme (enzyme·label) should be corrected for saturation by the ligand. If the K_d for the formation of the E·ligand complex is such that the complex is only partially saturated, then $f = $ [E·label·ligand]/[ligand]. In the cases where the label is a metal ion then saturation of the E·M·ligand complex must also occur or be corrected. The formation of binary M·ligand complexes must be minimized or corrected. The value for n, the mole fraction of M in the E·M·ligand complex, can be calculated from known dissociation constants or by a measure of $1/fT_{ip}$ under analogous conditions at three different values of ω_I (Nowak 1981).

If the values for $1/fT_{1p}$ and $1/ft_{2p}$ can be correctly determined and $\Delta\omega$ for E·M·L is negligible, evaluation of these parameters must be made. If $1/fT_{ip}$ is in fast exchange, $T_{iM} \gg \tau_m$, then

$$\frac{1}{fT_{1p}} = \frac{q}{T_{1M}} = \frac{1}{T_{1M}} \tag{12}$$

$$\frac{1}{fT_{2p}} = \frac{q}{T_{2M}} = \frac{1}{T_{2M}} \tag{13}$$

An evaluation of q, the number of ligands binding at the paramagnetic label site, can either be made by direct binding studies; or, in most cases, q is simply one.

These relationships are somewhat simplified by the assumption that outer sphere effects are negligible. These effects occur when ligands in solution approach the paramagnetic center but do not bind at the normal binding site (which may already be occupied). The time of interaction and the longer dipolar distance for these outer sphere ligands results in a small, usually insignificant effect. The measured relaxation rates can then be related to the structure of the ligand on the enzyme relative to the paramagnetic center. This information can be obtained from the Solomon–Bloembergen relationships (Solomon 1955, Bloembergen 1957). These relationships relate the dipolar (through space) and scalar (through chemical bonds) contributions of the paramagnetic centers to the nuclear relaxation rates:

$$\frac{1}{T_{1M}} + \frac{2S(S+1)\gamma_I^2 g^2 \beta^2}{15r^6} \left(\frac{3\tau_c}{1+\omega_I^2\tau_c^2} + \frac{7\tau_c}{1+\omega_S^2\tau_c^2} \right) +$$

$$\frac{2S(S+1)}{3} \left(\frac{A}{\hbar} \right)^2 \left(\frac{\tau_e}{1+\omega_S^2\tau_e^2} \right) \tag{14}$$

$$\frac{1}{T_{2M}} = \frac{S(S+1)\gamma_I^2 g^2 \beta^2}{15} \left(4\tau_c + \frac{3\tau_c}{1+\omega_I^2\tau_c^2} + \frac{13\tau_c}{1+\omega_S^2\tau_c^2} \right) +$$

$$\frac{S(S+1)}{3} \left(\frac{A}{\hbar} \right)^2 \left(\tau_e + \frac{\tau_e}{1+\omega_S^2\tau_e^2} \right) \tag{15}$$

The term S is the electron spin quantum number; γ_I is the magnetogyric (gyromagnetic) ratio of the nuclear spin; g is the electronic 'g' factor; β is the Bohr magneton; ω_I and ω_S are the Larmor angular precession frequencies for the nuclear and the electron spins respectively ($\omega_S = 57\,\omega_I$); r is the ion (spin–label electron)–nucleus distance; A is the hyperfine coupling constant; $\hbar$ is Planck's constant divided by π; and τ_c and τ_r are the correlation times for the dipolar and the scalar interactions respectively. The first term in each equation describes the dipolar interaction, and the second term describes the scalar interaction. Scalar interactions are caused by electron density from the electron spin at the nucleus under observation. Nuclei on ligands not directly bonded to the probe (metal) ions are expected to have no scalar effect, as is also the case for some nuclei such as ^1H. Scalar effects on $1/T_{2M}$ for ^{31}P nuclei in second sphere complexes appear to be significant in some cases, however (Lee & Nowak 1984).

Upon perusal of the Solomon–Bloembergen equations, it is clear that the correlation time functions for T_{1M} and T_{2M} are different, and therefore their frequency and temperature behavior should differ. For T_{1M}, scalar interactions are very small and can usually be ignored. Values for S, g and τ_s, the electron relaxation time for the paramagnet for a variety of possible paramagnetic probes, are listed in Table 10.1. The τ_s may be an important component of τ_c in eqs (14 and (15).

Table 10.1

Electron spin quantum numbers and approximate values of isotropic g factors and electron spin relaxation times for some paramagnetic probes.

Probe	S	$g^{\dagger}$	$\tau_s{}^{\ddagger}$ sec
Nitroxide (N—O)	$\frac{1}{2}$	2.0	10^{-8}
Cu(II)	$\frac{1}{2}$	2.0–2.5	10^{-10}
Fe(III), low spin	$\frac{1}{2}$	2–6	10^{-12}
V(IV)	$\frac{1}{2}$	2.0	10^{-9}
Ni(II)	1	2.0–2.9	10^{-13}
Co(II)	$\frac{3}{2}$	$2.1-2.8$	10^{-13}
Cr(III)	$\frac{3}{2}$	2.0	10^{-10}
Fe(II)	2	2.1–2.3	10^{-11}
Mn(II)	$\frac{5}{2}$	2.0	10^{-9}
Fe(III), high spin	$\frac{5}{2}$	2.0	10^{-10}
Gd(III)	$\frac{7}{2}$	2.0	10^{-10}

† These are approximate values that may be used in the Solomon–Bloembergen interactions. For lanthanide ions the electron angular momentum J is necessary. The g factor may not correspond to the true isotropic factor.
‡ The electron spin longitudinal relaxation time is dependent upon ligands coordinated to the ion, frequency, and temperature. These values are approximate for the spin in question.

Other constants used in the equation are $\beta = 9.284 \times 10^{-21}$ erg Gauss^{-1} and $\hbar = 1.055 \times 10^{-27}$ erg sec. Although the choice of probes may be dictated by the nature of the enzyme, its chemical, physical, and biophysical properties, MnZ(II) would, given a choice, be the probe to be used. This choice is predicated by its ability to substitute for a number of physiological metal ions (Mg(II), Ca(II), Zn(II)...), its labile hydration sphere (τ_m for Mn(II)(H$_2$O)$_6$ ~3 $\times 10^{-8}$s), large electron spin quantum number $(\frac{5}{2})$, normally isotropic behavior, and long electron spin relaxation time, τ_s.

In a study of Mn(II)-^{1}H interactions where $\gamma = 2.675 \times 10^4$ rad Gauss^{-1}

$\sec^{-1}$, $g = 2$ and $S = \frac{5}{2}$, and hyperfine coupling is negligible, the equations for $1/T_{1M}$ and $1/T_{2M}$ can be simplified as

$$\frac{1}{T_{1M}} = \frac{2.878 \times 10^{-31}}{r^6} f(\tau_c) \tag{16}$$

$$\frac{1}{T_{2M}} = \frac{1.439 \times 10^{-31}}{r^6} f'(\tau_c) \tag{17}$$

where $f(\tau_c)$ and $f'(\tau_c)$ are the correlation time functions for T_1 and T_2 relaxation respectively. A determination of τ_c can allow the calculation of r for the ^{1}H to Mn(II) in the E·Mn·ligand complex. The assumption of negligible hyperfine coupling for $1/T_{1M}$ usually holds, but with some nuclei, specifically ^{31}P of phosphates, this is not the case for $1/T_{2M}$.

Several methods may be used to estimate τ_c. Discussions concerning the rigor of such estimates are detailed elsewhere (Dwek 1973, Mildvan & Gupta 1978, Nowak 1981).

With a value for τ_c, r can thus be calculated. Choosing Mn(II) as an example, from a measurement of T_{1M} and an estimate of τ_c, r, in Angstroms, can be calculated:

$$r = X[T_{1M}f(\tau_c)]^{1/6} . \tag{18}$$

T_{1M} comes from direct measurements, and $f(\tau_c)$ is calculated at ω_I after an estimate of τ_c. The value X is a collection of constants giving $X = 812$ for ^{1}H; 796 for ^{19}F; 601 for ^{31}P; or 512 for ^{13}C. Because of the sixth-power dependence of r on T_{1M} the absolute value of r is reasonably insensitive to minor errors in the estimation of $f(\tau_c)$ or T_{1M}. On the other hand, small differences in r for different nuclei within the ligand give rise to large differences T_{1M}. Thus the method is very sensitive to small changes in the structure of the ligand on the enzyme. From a determination of the various values of r in the enzyme–label–ligand ternary complex the structure of the ligand relative to label can be determined.

In cases, for example, where a bisubstrate enzyme is apparently labeled at site 2 to yield inactive enzyme, the effect of this modification on the binding (K_d) and structure of substrate 1 at the catalytic site can be investigated.

In some cases where one nucleus of a ligand is very close to the paramagnetic center compared to other nuclei measured, the relaxation may be so efficient that the nucleus may be in slow exchange ($T_{2M} \ll \tau_m$) ($1/fT_{2p} = 1/\tau_m$). If this is the case then a temperature-dependence of $1/fT_{2p}$ will give a value for k_{off} and for the energy of activation, E_{act} for the ligand exchange process. In this case the structure of the ligand at the catalytic site (from $1/T_{1M}$), its exchange rate, and the energy barrier for this exchange process, can be obtained and compared with these parameters for the unmodified enzyme. In the case where the exchange process is simple, and

$$K_d = \frac{k_{off}}{k_{on}} \qquad (19)$$

for ligand binding is known, values of k_{on} can also be estimated.

WATER RELAXATION RATE PROCESSES

A rapid and sensitive method of measuring ligand–enzyme interactions, where the enzyme system is appropriate, is to measure the effect of ligand binding on the solvent (^1H of H_2O). This method requires a paramagnetic probe that can affect the longitudinal relaxation rate of the solvent. The probe elicits an effect on the proton longitudinal relaxation rate (PRR or PRE) to give a proton relaxation rate enhancement. If the enhancement effects are sensitive to ligand binding, then studying the environment around the probe can yield important thermodynamic and structural information.

Although a number of probes can be used for these studies, Mn(II) will again be chosen as an example because of its physical–chemical properties and its usefulness in many cases. The interaction of the solvent with the paramagnetic probe increases the relaxation rates of the ^1H's of H_2O by dipole–dipole interactions as discussed in the previous section. Such studies are usually performed measuring only T_1 values and, as will be seen, lower values of ω_I (15–40 MHz) are preferable to higher frequencies (100–400 MHz). Dedicated, low-resolution pulsed instruments have been designed especially for such studies.

In free solution the interaction of the unpaired electrons of the metal ions or of the nitroxide with the ^1H nuclei can be normalized as shown in eq. (9). For metal ions the number of protons is twice the hydration number, q, (2×6 for Mn(II); 2×8–9 for Gd(III). Hydration may be less clear for organic nitroxides. In solution at room temperature rapid exchange conditions prevail and the correlation time is often τ_r. For $Mn(H_2O)_6$, τ_r is approximately 2.9×10^{-11}s and with $\tau_s \sim 10^{-9}$s and the residence time for Mn-bound water, $\tau_m \sim 2 \times 10^{-7}$s, τ_c is τ_r. For some metals where τ_s is short (for Fe(II), $\tau_s \sim 10^{-11}$s) τ_c is determined by τ_r and τ_s. When the metal binds to an enzyme, at least two phenomena occur: q decreases, decreasing the value for $1/fT_{1p}$ and τ_r increases. The increase in τ_r, if τ_r modulates relaxation, increases $1/fT_{1p}$. For probes with long τ_s values, an increase in τ_r to the rotational correlation time of the enzyme ($\sim 10^{-9} - 10^{-7}$s) results in a substantial increase in τ_c to the point where τ_c may be dominated by processes other than τ_r. Thus the $1/fT_{1p}$ for ^1H of H_2O by the paramagnetic species is enhanced by some factor ε^*. The observed enhancement can be quantitated by comparing the paramagnetic effect of the label in the presence of enzyme, designated by the asterisk (*), to that in its absence:

$$\varepsilon^* = \frac{\left(\dfrac{1}{T_{1p}}\right)^*}{\left(\dfrac{1}{T_{1p}}\right)} = \frac{\left(\dfrac{1}{T_{1,\text{obs}}} - \dfrac{1}{T_{1,0}}\right)^*}{\left(\dfrac{1}{T_{1,\text{obs}}} - \dfrac{1}{T_{1,0}}\right)} \tag{20}$$

The control used for the denominator term is simply the paramagnetic effect of the probe on the PRR measured in the absence of enzyme.

The observed relaxation rate is the sum of the paramagnetic effects due to free species and bound species:

$$\left(\frac{1}{T_{1p}}\right)^* = \left(\frac{f\,q}{T_{1M} + \tau_m}\right)_f + \left(\frac{f\,q}{T_{1M} + \tau_m}\right)_b \tag{21}$$

The term designated subscript b describes the normalization factor f for the concentration of free paramagnetic species ($[p]f$) and the concentration of water (55.5M) ($f = [p]f/55.5$); q, the hydration number for free metal ion (nitroxide); T_{1M} for the ^{1}H of H_2O bound to the free probe; and the lifetime, τ_m of the complexes. The term designated subscript b describes the same parameters for the species which is enzyme bound and is primarily responsible for enhancement. This effect, or the enhancement, can be quantitated by relating the free and bound paramagnetic species $[M]_f$ and $[M]_b$ respectively:

$$\varepsilon^* = \frac{[M]_f}{[M]_T}\,\varepsilon_f + \frac{[M]_b}{[M]_T}\,\varepsilon_b \tag{22}$$

The enhancement of free species, ε_f, is defined as unity, and the binary enhancement (of the enzyme–label complex), ε_b is characteristic of the complex being studied. The preceding equation relates the mole fraction of each species. Direct binding studies of the probe can be used to evaluate ε_b. The value for ε_b is a reflection of the environment about the bound probe, and contains information concerning q, τ_m, and τ_c for the H_2O at the probe. A comparison of stoichiometry, K_d, and ε_b for metal binding to modified and unmodified enzyme can relate the effect of modification to the environment of the activator. If the stoichiometry of an enzyme–M complex is known or assumed to be 1:1, a titration of ε^* versus $[M]$ can yield K_d and ε_b.

An actual quantitation of q for H_2O in the enzyme–label complex can be obtained by a determination of τ_c for the complex. Several methods to determine τ_c for Mn(II)-H$_2$O interactions in binary enzyme–Mn complexes have been attempted to evaluate these parameters. Such approaches are not very simple, and the evaluation of additional physical–chemical parameters and several assumptions are required. Reasonable approximations to these parameters can be made, but the evaluation of the actual hydration number by such studies should be taken with some amount of skepticism (Burton *et al.* 1979). One important detail obtained from frequency-dependent studies of $(1/T_{1p})$ is that for enzyme-bound Mn(II), the value for τ_c is usually frequency-dependent, showing that τ_c must be at least partially determined by τ_s in those cases. A fit of the data to equations (9) and (14) would suggest that fT_{1p} is linear with $(\omega_I)2$ if τ_c is constant. Lower frequency measurements should yield greater paramagnetic effects on T_1.

This approach to the study of ligand interactions with enzymes can also be used when paramagnetic ions bind only to enzyme–ligand complexes (i.e. creatine kinase where Mn(II) binds to creatine kinase ATP but not to creatine kinase). A similar evaluation of the data can be made.

The addition of a ligand (substrate or allosteric modifier) to the over enzyme–label complex can result in a perturbation of one or several parameters which influence relaxation. The bound ligand can change metal binding (K_d and/or n), resulting in a change in [M]$_b$ or can result in changes in q, τ_m, or τ_c. A change in τ_c can affect T_{1M}. Regardless of the reason for the perturbation in $1/T_{1p}$, a change in ε^* upon addition of ligand may be obtained. Such a titration can result in a determination of the dissociation constant of the ligand from the enzyme complex and a value for the enhancement of the ternary enzyme–label–ligand complex, ε_t. The equation for observed enhancement now becomes

$$\varepsilon_{obs} = \frac{[M]_f}{[M]_T}(1) + \frac{[E \cdot M]}{[M]_T}\varepsilon_b + \frac{[E \cdot M \cdot L]}{[M]_T}\varepsilon_T \qquad (23)$$

In the case of metal complexes where the ligand also competes with the enzyme for metal binding, a term [M-L]/[M]$_T$(ε_a), must also be considered. A value for the enhancement of the metal–ligand complex, ε_aa, can be evaluated independently. In a general case when all possible equilibria may be present in such a titration experiment, the following complexes, their dissociation constants, and enhancement values must be considered:

$$K_D = \frac{[E][M]}{[EM]}\varepsilon_b \qquad (24)$$

$$K_1 = \frac{[M][L]}{[ML]}\varepsilon_a \qquad (25)$$

$$K_2 = \frac{[E][M-L]}{[EML]} \varepsilon_T \tag{26}$$

$$K_3 = \frac{[EM][L]}{[EML]} \varepsilon_T \tag{27}$$

$$K_A = \frac{[EL][M]}{[EML]} \varepsilon_T \tag{28}$$

$$K_S = \frac{[E][L]}{[EL]} \tag{29}$$

From thermodynamics

$$K_1 K_2 = K_D K_3 = K_A K_S . \tag{30}$$

In a simple case, the addition of a ligand to an enzyme–label which results in a change in ε^* indicates the formation of an enzyme–label–ligand ternary complex. Values for ε_T can be either greater than ε_b or less than ε_b; no change in ε^* may result from failure of ligand to bind; no change in physical parameters affecting ε^*; or fortuitous changes in $1/T_{1p}$ that result in no observed change. A change in ε^* with concentration of ligand can be either graphically evaluated to yield K_d and ε_T, or a fit to the data can be attempted. A computational analysis of ε^* vs [ligand], considering all possible equilibria, has been developed (Reed et al. 1970) and is the more rigorous treatment. An evaluation of K_3 and ε_T reflects any change of enzyme modification on ligand binding and on any environmental change about the probe induced by the ligand respectively. Such changes can be compared to those observed in the native or non-derivatized enzyme. This method can be a powerful yet simple tool to evaluate the effects of enzyme modification on ligand binding.

Analogous to PRR measurements, the environment of the catalytic site can also be studied by observing the relaxation ($1/T_2$ from the line width measurements) of inorganic anions which serve as part of the millieu. The interaction of an anion such as Cl^- using ^{35}Cl NMR with paramagnetic centers can elicit a paramagnetic effect analogous to effects observed using PRR. On the other hand, the interaction with diamagnetic centers can also give rise to substantial quadrupolar relaxations resulting in T_2 effects. The observation of such an effect will demonstrate that the diamagnetic center, i.e. Zn(II), has access to the solvent. The addition of ligands influences the metal center. Another variation is the use of the more sensitive ^{19}F nucleus of F^- if this ion is innocuous for the enzyme under investigation. The F^- interaction with the Cu(II) center of galactose oxidase was studied by relaxation rate measurements (Marwedel et al. 1975). The limitation of ^{19}F NMR is that F^- is an inhibitor for a number of enzymes and may not be a good anion in every case. The study of anion relaxation has important potential in the study of environment effects and ligand effects on the active-site environment.

CATION NMR

An alternative to studying the enzyme directly by NMR is to observe a portion of the enzyme, preferably the active-site, where good resolution of the important functional groups can be observed. One example where such a study is possible is in the case of metal utilizing enzymes where the spectra of one of several metals can be observed. These metal ions often play a key role in the catalytic processes. There is a potential for the use of the 'physiologically common' divalent cations ^{25}Mg and ^{43}Ca; a review of the attempted applications contains some examples (Forsén & Lindman 1981). Most of the studies have been with ^{43}Ca which is somewhat easier to study. The primary drawback in these studies is the very low sensitivity of these nuclei. The high concentration of these cations necessary to observe the resonances of bound cations are limited by the solubility of the proteins.

Applications of monovalent cations ^{7}Li, ^{23}Na, ^{39}K or ^{205}Tl are also quite possible. Because of much weaker binding of the monovalent cations to enzymes, the effects, including ligand–induced cation perturbations, can also be observed and perhaps quantitated.

A broader application of ^{113}Cd NMR has been made. Although Cd(II) is a toxic metal ion, it has been found to substitute for Zn(II) or Mg(II) in several enzyme systems. Its chemical shift is sensitive to the nature of its ligands, and a change in ligand environment (e.g. addition of substrates) is reflected in a change in chemical shift (Armitage & Otvos 1982). The nucleus can be used to study chemical exchange between metal sites, multimetal sites, and the interactions of ligands to the enzyme–bound metal. The low sensitivity and low resonance frequency keeps this metal from being routinely used. It can prove to be quite useful, however, in the cases where plenty of enzyme that is quite soluble can be obtained.

The goal of this chapter is to stimulate the reader to use NMR spectroscopy as a tool in conjunction with protein modification studies to obtain detailed information on enzyme active-sites and their functions. Although the applications of NMR to such problems are not treated exhaustively, the potential information to be garnered and the limitations are discussed. As the appropriate NMR applications are chosen for the specific enzymes in question, additional information can be found in the current literature to assist in the fundamental understanding. There are numerous fascinating problems awaiting our exploration.

GLOSSARY OF TERMS

A	hyperfine coupling constant
g	electronic 'g' factor
H_o	intensity of the stationary magnetic field

ℏ	Planck's constant divided by 2π
I	total nuclear spin quantum number
J	Scalar spin–spin coupling constant
K	equilibrium constant
k	rate constant
M_x, M_y, M_z	components of the macroscopic magnetic moment along the x, y, and z axes, respectively
P	angular momentum
q	number of ligands bound in the sphere of influence of the paramagnetic species
r	internuclear distance
S	total electron spin quantum number
T	absolute temperature
T_1	longitudinal (spin–lattice) relaxation time
T_2	transverse (spin–spin) relaxation time
T_{1M}, T_{2M}	relaxation times for nuclei which are in the sphere of influence of the paramagnetic species
β	Bohr magneton
τ	gyromagnetic (magnetogyric) ratio
δ	chemical shift (ppm)
ε	water proton spin lattice relaxation rate enhancement
μ	magnetic moment of an individual spin
ν	frequency (Hz)
τ_c	correlation time
τ_e	hyperfine correlation time
τ_m	lifetime of a nucleus in the coordination sphere of a paramagnetic species
τ_r	rotational correlation time
τ_s	electron spin relaxation time
ω_I	nuclear resonance frequency ($2\pi\nu$)
ω_s	electron resonance frequency

REFERENCES

Abragam, A. (1973) *The Principles of Nuclear Magnetism*, Oxford University Press, Oxford.

Armitage, I.A. & Otvos, J.D. (1982) In: *Biological Magnetic Resonance* (Berliner, L.J. & Reuben, J. eds) Vol 4, p. 79, Plenum Press, New York.

Becker, E.R. (1980) *High Resolution NMR*, 2nd edition, Academic Press, New York.

Berliner, L.J. (1976) *Spin Labeling, Theory and Application*, Academic Press, New York.

Bloch, F. (1946) *Phys. Rev.* **70** 460.

Bloembergen, N. (1957) *J. Chem. Phys.* **27** 572.

Burton, D.R. Forsén, S., Karlstrom, G. & Dwek, R.A. (1979) *Progress in NMR Spectroscopy* **13** 1.

Cleland, W.W. & Mildvan, A.S. (1979) *Adv. Inorg. Biochem.* **1** 63.

Cohn, M. & Townsend, J. (1954) *Nature (London)* **173** 1090.

Dwek, R.A. (1973) *NMR in Biochemistry*, Oxford University Press, Oxford.

Forsén, S. & Lindman, B. (1981) *Methods in Biochemical Analysis* **27** 289.

Gerig, J.T. (1981) *Biological Magnetic Resonance* (Berliner, L.J. & Reuben, J. eds) Vol 1, p. 139, Plenum Press, New York.

Hunkapiller, M.W., Smallcombe, S.H., Whitaker, D.R. & Richards, J.H. (1973) *Biochemistry* **12** 4732.

James, T.L. (1975) *Nuclear Magnetic Resonance in Biochemistry*, Academic Press, New York.

Jardetzky, O. & Roberts, G.C.K. (1981) *NMR in Molecular Biology*, Academic Press, New York.

Kaptein, R. (1982) In: *Biological Magnetic Resonance* (Berliner, L.J. & Reuben, J. eds) Vol 4, p. 145, Plenum Press, New York.

Lee, M.-H. & Nowak, T. (1984) *Biochemistry* **23** 6506

Luz, Z. & Meiboom, S. (1964) *J. Chem. Phys.* **40** 2686.

Malthouse, J.P.G. (1986) *Prog. in Nuclear Magnetic Resonance Spectros.* **18** 1.

Marwedel, B.J., Kurland, R.J., Kasman, D.J. & Ettinger, M.J. (1975) *Biochem. Biophys. Res. Comm.* **63** 773.

Meadows, D.H. & Jardetzky, O. (1986) *Proc. Natl. Acad. Sci. USA* **61** 406.

Meshitsuka, S., Smith, G.M. & Mildvan, A.S. (1981) *J. Biol. Chem.* **256** 4460.

Mildvan, A.S. & Gupta, R. (1978) *Methods Enzymol.* **49** 322.

Nowak, T. (1981) In: *Spectroscopy in Biochemistry* (Bell, J.E. ed.) Vol 2, p. 109, CRC Press, Boca Raton, FL.

Reed, G.H., Cohn, M. & O'Sullivan, W.J. (1970) *J.Biol. Chem.* **245** 6547.

Roberts, G.C.K. & Jardetzky, O. (1970) *Advances Protein Chem.* **24** 447.

Schowen, R.L. (1972) *Prog. Phys Org. Chem.* **9** 275.

Solomon, I. (1955) *Phys. Rev.* **99** 559.

Swift, T.J. & Connick, R.E. (1962) *J. Chem Phys.* **37** 307.

Sykes, B.D. & Weiner, J.H. (1980) *Magnetic Resonance in Biology* (Cohen, J.S. ed.) Vol 1, p. 171, John Wiley, New York.

Turner, C.J. (1984) *Progress in Nuclear Magnetic Resonance Spectroscopy* **16** 311.

Wüthrich, K., Wider, G., Wagner, G. & Braun, W. (1982) *J. Mol. Biol.* **155** 311.

11

Monoclonal antibodies as probes for structural and functional features of proteins

Dr John E. Wilson, Biochemistry Department and the Neuroscience Program, Michigan State University, East Lansing, MI 48824, USA

One strategy that is frequently exploited in establishing structure–function relationships for proteins is to modify a specific structural feature and look for an effect on function. If the latter is found, it implies that the modified structural feature is involved, either directly or indirectly, in that function. Certainly most common would be techniques by which specific amino acid residues or ligand binding sites are modified by chemical means, techniques discussed elsewhere in this book. In the present chapter, we consider the utility of monoclonal antibodies as probes for specific structural and functional features of protein molecules. We will not attempt to review the already extensive literature on this subject, but will draw particularly on our own work using monoclonal antibodies against rat brain hexokinase (Finney et al. 1984, Polakis & Wilson 1984, 1985, Wilson & Smith 1985, Ureta et al. 1986) as well as other examples selected from the recent literature.

POLYCLONAL VS MONOCLONAL ANTIBODIES

Injection of an animal with an immunogenic protein sets off a complex and still poorly understood sequence of events that culminates in stimulation of a population of cells, each of which gives rise to a clonal cell line producing a single type of antibody directed against a specific structural feature (epitope) on the protein. Many such cell lines are generated, since a typical protein contains numerous antigenic sites capable of eliciting a response. Hence, the resulting polyclonal antibodies are extremely heterogeneous, and represent a collection of antibody molecules capable of simultaneously interacting with several different epitopes on the protein. Although such a diversity of potential interactions does not prevent, and may enhance, the use of polyclonal antisera in other applications (e.g., immunoprecipitation or immunohistochemistry), it obviously precludes their use as probes for specific structural features which may be linked to function by the strategy noted above.†

The development of methods for preparing monoclonal antibodies (Kohler & Milstein 1975) has made it possible to obtain virtually unlimited amounts of truly homogeneous preparations of antibodies directed against specific epitopes. Briefly, an animal (usually a mouse or rat) is immunized with the protein, and, subsequently, lymphocytes are prepared from the spleen or lymph nodes. These cells possess the capability for producing antibodies but are not capable of being propagated in culture. However, if these cells are fused with a suitable myeloma cell line, the resulting 'hybridomas' possess two highly desirable properties: they are capable of producing antibodies, a capability inherited from their lymphocyte pre-cursor, and they can be propagated indefinitely in culture, a characteristic conferred by the myeloma. Using appropriate culture conditions, it is readily possible to select for hybridomas, i.e. the culture medium is such that only the hybridomas grow while residual (unfused) lymphocyte and mye-loma cells die off. Since the original lymphocyte population was hetero-geneous (i.e. the immunized animal was capable of making a diversity of antibodies against both the injected protein as well as other antigens it had encountered), it is evident that hybridomas resulting from the fusion will also represent a heterogeneous population capable of secreting a variety of antibodies, only some of which will be directed against the antigen of interest.‡ Hence the next step involved in monoclonal antibody production consists of 'screening' the hybridomas for those producing antibodies against the protein of interest, then cloning the desired hybridomas to yield a

---

† In principle, it is possible to obtain selected subpopulations of antibodies from polyclonal antiserum by use of appropriate adsorption methods. For example, it may be possible to isolate antibodies directed against a specific region of the protein's structure by employing an affinity column containing a proteolytic fragment derived from the region of interest as the immobilized ligand. However, it is highly unlikely that such techniques would ever yield truly homogenous populations of antibodies. First of all, there may be multiple, perhaps overlapping, epitopes within any given polypeptide segment. Secondly, even if all the antibodies obtained by such a method recognized exactly the same epitope, it is still conceivable that they may have originated from different clonal cell lines, i.e. despite their recognition of a common epitope, they are chemically distinct antibody molecules.

‡ Lo *et al.* (1984) have described a method for selectively fusing myelomas with spleen cells capable of generating antibodies against the desired antigen, i.e. virtually all of the resulting hybridomas will secrete antibodies against the protein of interest. Though this promises to greatly increase the efficiency of the hybridoma-generation process, the method has not yet come into common use.

cell line secreting a homogeneous antibody population directed at a specific epitope.†

ANTIGENIC SITES ON PROTEINS

If our objective is to secure antibodies against specific structural features of a protein, it is reasonable to ask what types of structural features might be effective at eliciting an immune response.

Segmental vs. conformational epitopes

It is common to divide epitopes into two major categories (Berzofsky 1985): 'segmental' (or 'continuous', or 'sequential') and 'assembled topographic' (or 'conformational') sites. A segmental epitope may be defined as one in which all of the structural features required for recognition by the antibody reside in a single short (approximately 5–8 amino acid residues) segment of the polypeptide chain. In contrast, an epitope of the assembled topographic type would include amino acid residues that may be quite distant within the amino acid sequence but are brought into proximity during folding of the protein. It is apparent that recognition by antibodies directed against assembled topographic epitopes would depend greatly on the preservation of requisite secondary and/or tertiary structural features (i.e. denatured proteins would show greatly decreased or totally abolished immunoreactivity), while epitopes of the segmental type should be reactive, if accesssible, in both native and denatured forms of the enzyme. It also follows that reactivity with antibodies directed against epitopes of the assembled topographic type might be influenced by perturbations of the tertiary structure, such as conformational changes induced by ligand binding, whereas this seems less likely with segmental epitopes.

Since even limited segments of polypeptide could exist in many conformations, and it is unlikely that all are effectively recognized by the antibody, conformation must be a factor in recognition even of segmental epitopes (Berzofsy 1985). Nevertheless, the distinction between segmental and assembled topographic epitopes is still useful since clearly the required conformational features are much more restrictive (in terms of overall secondary and tertiary structure) with the latter type.

---

† It is worth noting the practical implications of this screening process. In contrast to the case with polyclonal antiserum, it is *not* necessary to immunize the animal with pure antigen in order to obtain monospecific antibodies. Even if injection of impure antigen results in production of antibodies against impurities, the corresponding hybridomas are discarded in the screening process, which is selective for antibodies against the antigen of interest. Although, in principle, crude preparations of antigen could be used as immunogen, in practice it is usually advisable to use at least a reasonably pure preparation so as to focus the immunological response on the antigen of interest.

It will also be apparent from the above comments that monoclonal antibodies provide a valuable approach for purification of antigens difficult to obtain by conventional methods. Thus, an impure preparation of the antigen can be used to obtain a monoclonal antibody, which can then be used to purify the antigen by immunoaffinity methods.

Dependence of antibody production on conformational state of the antigen
It generally seems to be the case that resulting polyclonal antibodies react
best with the form of the antigen used for immunization (Jemmerson &
Paterson 1986). Thus if the native enzyme is used as immunogen, one may
expect a population of antibodies directed primarily at epitopes on the
native enzyme. The majority of the antibodies induced by immunization
with a native protein appear to be directed against assembled topographic
sites (Berzofsky 1985). In contrast, if denatured protein is used as immunogen one can expect antibodies more reactive with the unfolded polypeptide
chain. Using aldolase denatured by heating in the presence of sodium
dodecylsulfate and 2-mercaptoethanol as antigen, Reznick *et al.* (1985)
obtained a polyclonal antiserum directed exclusively against the denatured
enzyme. Presumably, antibodies obtained with the native enzyme as immunogen but recognizing segmental epitopes would also be reactive with the
denatured enzyme. Antibodies against these same epitopes may be
expected with denatured enzyme as antigen, as well as additional antibodies
directed against epitopes that became exposed owing to denaturation.

Considering that the adjuvants commonly used as vehicles for injection
of antigen would hardly be selected as the solvents of choice for maintaining
native protein structure, and considering all the trauma that must assault a
protein's structure as it proceeds from being injected to actually evoking an
immune response, this dependence of the immunological result on the
structural status of the antigen seems surprising. And in the case of
monoclonal antibodies, it is clear that immunization with 'native' (meaning,
not *purposely* denatured) antigen does indeed generate a spectrum of
antibodies, at least some of which react solely with denatured forms of the
antigen (Mierendorf & Dimond 1983, Djavadi-Ohaniance *et al.* 1984,
Finney *et al.* 1984, Sams *et al.* 1985, Dunn *et al.* 1985, Wilson & Smith 1985,
Vaidya *et al.* 1985) and appear to recognize epitopes not normally exposed in
the native structure. Though antibodies of this type will not be useful in
establishing structure–function relationships by the strategy mentioned at
the start of this chapter, they can certainly be useful in other contexts and
may aid in the establishment of structure–function relationships by alternative methods (Polakis & Wilson 1984, 1985, Ureta *et al.* 1986).

Distribution of epitopes on the molecular surface
Even though it is clear that immunization with native protein can give rise to
antibodies specific for the denatured form, it is also clear that many of the
resulting monoclonal antibodies react preferentially if not exclusively with
the native form used for immunization (e.g. Finney *et al.* 1984, Dunn *et al.*
1985), and are likely to be directed against assembled topographic sites
(Berzofsky 1985). There has been considerable discussion of what dictates
the antigenicity of a particular site on a protein. In the case of the native
structure, it is evident that any antigenic site must be accessible at the surface
of the folded structure. Thus there has been considerable evidence to
indicate that the relative hydrophilicity of particular segments within the
polypeptide is related to antigenicity, the rationale being that the more

hydrophilic regions are more likely to be located at the protein surface and are hence capable of being antigenic (Hopp & Woods 1981). More recently, attention has been focused on the relative mobility of particular polypeptide segments at the surface of the molecule, with greater conformational flexibility being correlated with greater antigenicity (Moore & Williams 1980, Tainer et al. 1984, Westhof et al. 1984).

Although the relative antigenicity of various surface regions may differ, the current view seems to be that the entire surface is potentially antigenic (Berzofsky 1985, Jemmerson & Paterson 1986). Assuming that this were correct, it might be anticipated that the relative distribution of epitopes within the overall structure of the protein would be more-or-less random (some comments on epitope mapping techniques will be given later). While this may sometimes be the case, a remarkable number of studies have indicated 'clustering' of epitopes in rather restricted regions of the molecule. Epitopes for seven monoclonal antibodies reactive with the native form of brain hexokinase were mapped to a relatively limited area in the N-terminal region of the molecule (Wilson & Smith 1985). Weldon & Taylor (1985) noted that the C-terminal two-thirds of the cAMP-dependent protein kinase II was 'remarkably nonantigenic', while the results of Minitz & Brimijoin (1985) suggested that epitopes recognized by monoclonal antibodies against rabbit brain acetylcholinesterase 'were not widely dispersed on the antigen surface', and similar clustering has been found for epitopes recognized by monoclonal antibodies against other proteins (e.g. Crawford et al. 1982, Lillenhoj et al. 1982, Vartio et al. 1982, Sams et al. 1985).

What would make such large regions of a molecule so non-antigenic? One possible explanation might be that the structure in these regions is highly conserved (Vartio et al. 1982). At least with polyclonal antisera, it appears that the antibodies are largely directed against epitopes that are not present on the homologous protein of the immunized animal (Berzofsky 1985, Jemmerson & Paterson 1986). This 'self-tolerance', i.e. the failure to produce antibodies against epitopes found on both the injected antigen and the homologous host protein, is widely viewed as a protective mechanism, the breakdown of which can result in autoimmune disease. The occurrence of the latter demonstrates that the capacity to produce antibodies against 'self' is in fact present, but suppressed under normal conditions. But while 'self-tolerance' may serve a useful role in the case of circulating polyclonal antibodies, it is not evident why this biological raison d'etre should apply in the case of monoclonal antibodies which are produced in vitro. And while, in some cases (e.g. Rakonczay & Brimijoin 1986) a majority of the monoclonal antibodies examined have been found to be non-reactive with the homologous host protein, this is by no means generally true (e.g. Wolf et al. 1983, Ureta et al. 1986).

At least at present, 'self-tolerance' to highly conserved regions of homologous enzymes does not offer a particularly satisfactory explanation for the apparently weak antigenicity of such large regions of molecular surface. It seems that we have much to learn about the factors determining antigenicity.

MONOCLONAL ANTIBODIES AS PROBES FOR ESTABLISHING STRUCTURE–FUNCTION RELATIONSHIPS

Binding of an antibody in direct proximity to a catalytic site may be expected to affect activity, e.g. by blocking access of substrate to the site. Binding elsewhere on the molecule could also have an effect by inducing a conformational change (or preventing one) that influenced catalytic activity. Inhibition or activation by monoclonal antibodies has been seen with many enzymes (e.g. Wolf *et al*. 1983, Lad *et al*. 1984, Dunn *et al*. 1985, Hessova *et al*. 1985, Makino *et al*. 1985, Rockwell *et al*. 1985). If the location of the epitopes for activity–affecting monoclonal antibodies can be mapped within the enzyme structure, these antibodies represent a valuable tool for suggesting structure–function relationships. Thus, if the epitopes are clustered in a particular region of the molecule, it is apparent that that region is, either directly or indirectly, involved in catalytic activity. This may be particularly useful in assessing the functional roles of different subunits in multisubunit enzymes (e.g. Dunn *et al*. 1985, Rockwell *et al*. 1985), or in dissecting discrete functions within multifunctional proteins (e.g. Makino *et al*. 1985). Conversely, mapping the location of epitopes for monoclonal antibodies *not* having an effect on activity could tend to indicate those regions of the molecule that serve functions other than catalysis. For example, of seven monoclonal antibodies reactive with native rat brain hexokinase, none have any detectable effect on catalytic activity (Finney *et al*. 1984, Wilson & Smith 1985); and, as noted above, all bind to epitopes located in the N-terminal region of the molecule. These results suggest that the catalytic site lies elsewhere, i.e. toward the C-terminal region, and that is indeed the region of the molecule which is specifically labeled with reactive analogs of the substrates, ATP and glucose (Schirch *et al*. 1986). These results also suggest that some function other than catalysis may be associated with the N-terminal region of hexokinase. One such function is the specific interaction of this enzyme with the outer mitochondrial membrane, and epitopes for monoclonal antibodies capable of blocking this interaction have been mapped to a 10 000 mol. wt. 'binding domain' located at the extreme N-terminus of the molecule (Polakis & Wilson 1985, Wilson & Smith 1985).

As discussed above, there have been many cases in which monoclonal antibodies have been found to affect catalytic activity. Yet it is equally important to note that, overall, most monoclonal antibodies seem to bind without effect on catalytic effectiveness, e.g. seven monoclonal antibodies reactive with native rat brain hexokinase which do not affect catalytic properties of the enzyme (Finney *et al*. 1984, Wilson & Smith 1985). Of 90 monoclonal antibodies generated against rabbit skeletal muscle phosphorylase, only five had any effect on phosphorylase activity (Hessova *et al*. 1985), and similar lack of effect on catalytic activity has been seen with monoclonal antibodies against other enzymes (e.g. Dunn *et al*. 1985, Minitz & Brimijoin 1985, Santos *et al*. 1985, Weldon & Taylor 1985, Rakonczay & Brimijoin 1986). In terms of using monoclonal antibodies as probes for specific structural regions that may be related to function, these results are reassuring. They suggest that indirect effects on catalytic activity (e.g. due to conformational alteration resulting from binding of the antibody to an epitope remote from the catalytic site), and presumably other functions, will not be a common occurrence, though one must of course remain aware of this possibility. In other words, effects of monoclonal antibodies on function may often be attributed to their binding at a structural feature *directly* involved in the affected function. The selective inhibition of hexokinase binding to mitochondria only by monoclonal antibodies recognizing epitopes in the N-terminal 'binding domain' (Polakis & Wilson 1985, Wilson & Smith 1985) is an example.

EPITOPE MAPPING

The utility of monoclonal antibodies in establishing structure–function relationships is obviously dependent on a knowledge of the location of the epitopes within the structure, i.e. epitope mapping. A variety of approaches have been used for obtaining such information, but we certainly will not attempt to review them all. Generally speaking, these approaches can be divided into two major categories, which we will call 'competitive' and 'direct' epitope mapping.

Direct epitope mapping
We will define direct epitope mapping as a procedure by which the location of an epitope may be assigned to a particular segment of the molecule having a known location within the overall amino acid sequence. In some cases, with segmental epitopes this may be possible with a resolution on the order of a few amino acid residues. Most commonly, direct epitope mapping depends on an ability to cleave (e.g. by proteolysis or chemical means) the protein into fragments having a known orientation and location within the overall sequence. By determining the reactivity of monoclonal antibodies with the various cleavage fragments, the location of the epitope can be deduced. The immunoreactivity of the fragments may be assessed using ELISA (e.g. Djavadi-Ohaniance *et al.* 1984) or immunoblotting (e.g. Vartio *et al.* 1982, Polakis & Wilson 1985), or by determining the ability of the isolated fragments to compete with the intact protein for binding to monoclonal antibodies (e.g. Weldon *et al.* 1983). Recently, electron microscopic techniques have been used to directly visualize the location of epitopes (Lamy *et al.* 1985, Dixit *et al.* 1986) within the structure of the intact protein.

Since the most commonly used methods for direct epitope mapping involve cleavage and physical separation of distinct portions of the molecule, often under extremely denaturing conditions (e.g. SDS gel electrophoresis), it might be expected that these methods would be applicable only to segmental epitopes. While in the limit this must certainly be true, it does not appear to be generally so. Thus, several monoclonal antibodies that react with native rat brain hexokinase but not the denatured enzyme, and thus presumably recognize epitopes of the assembled topological type, show immunoreactivity after Western blotting of even relatively small proteolytic fragments of the enzyme (Polakis & Wilson 1984, 1985, Wilson & Smith 1985). Presumably there is some renaturation that occurs during the blotting process that permits restoration of detectable immunoreactivity. Deletion of SDS from the blotting buffer promotes this, while inclusion of SDS in the buffer facilitates subsequent detection of epitopes (presumably segmental) accessible only in the denatured enzyme (Polakis & Wilson 1984).

Competitive epitope mapping

We define competitive epitope mapping as a procedure in which the ability of two monoclonal antibodies to bind simultaneously to the enzyme is determined. Such procedures do not directly locate epitopes in an absolute sense, but rather can indicate their relative distribution within the overall structure. While competitive techniques could obviously be used in assessing whether the epitopes for two antibodies lie within a given proteolytic fragment (i.e. for direct epitope mapping), we will be most interested in their application to intact native enzyme.

From considerations of the size of an IgG molecule, it has been estimated that, if two antibodies (of the IgG class) bind simultaneously to a protein, their epitopes must be separated by a distance of at least 35 Å across the surface of the protein molecule (Tzartos *et al.* 1981). If two antibodies cannot bind simultaneously, this implies that their epitopes are separated by less than the requisite 35 Å, while mutual hindrance but not total exclusion of simultaneous binding implies marginal overlap of the epitopic regions.[†] A number of techniques have been used to assess the degree of competition in antibody binding (e.g. Conti-Tronconi *et al.* 1981, Crawford *et al.* 1982, Lillenhoj *et al.* 1982, Wilson & Smith 1984), and we have found acrylamide gradient gel electrophoresis (Lambin & Fine 1979) to be particularly convenient (Wilson & Smith 1984).

Correlation of direct and competitive epitope mapping results

While competitive epitope mapping techniques provide information about the relative distribution of epitopes on the enzyme surface, they must in turn be related to some fixed point (or points) of reference to be useful in an absolute sense. The latter can be provided by direct mapping techniques. Together with other structural information and knowledge of the effects of the monoclonal antibodies on various functions of the enzyme, a three-

---

† An alternative explanation of mutually exclusive or hindered binding might be that binding of one antibody induces a conformational change affecting binding of the other antibody, with the epitope for the latter not necessarily being in spatial proximity to that of the first antibody. While one should certainly be aware of this as a potential explanation for mutually exclusive or hindered binding, it does not appear to be a common problem. Using direct epitope mapping techniques, a number of studies have demonstrated that the epitopes for mutually competitive antibodies were in fact located in the same limited regions of the amino acid sequence (see Wilson & Smith 1984, and references therein; Wilson & Smith 1985).

dimensional model can be developed that relates structural, immunological, and functional features in the molecule. An example of such an exercise is the recent work with rat brain hexokinase (Wilson & Smith 1985).

Monoclonal antibodies: some other applications
The emphasis in the present chapter has been on the potential usefulness of monoclonal antibodies as probes for specific structural and functional features in proteins, and in establishing structure–function relationships. Once established, these can serve as the basis for other work of general interest to enzymologists. For example, it may be expected that structural features of primary importance to function would be widely conserved among homologous enzymes; comparative immunological studies with monoclonal antibodies offer a convenient approach for assessing the distribution of such structural features in the enzyme from different organisms (Ureta *et al.* 1986).

Holowka *et al.* (1985) have described an intriguing application of two monoclonal antibodies directed against epitopes of known spatial distribution within the IgE molecule. Fab fragments from these two monoclonal antibodies were labeled with suitable donor and acceptor molecules, and fluorescence resonance energy transfer measurements were used to assess the distance between the epitopes in the free IgE molecule compared to IgE bound to its receptor in membrane vesicles. The result was development of a model for the effects of receptor binding on the conformation of the IgE molecule. One could anticipate analogous applications of suitably tagged monoclonal antibodies having epitopes of defined spatial relationship on other proteins, e.g. for studying the effects of ligand induced conformational changes on spatial relationship of the epitopes.

ACKNOWLEDGEMENTS
Our continuing efforts to utilize monoclonal antibodies in structure–function studies with brain hexokinase have been supported by NIH Grant NS-09910 and NSF Grant BNS-8320282. It is a pleasure to acknowledge the participation of Ken Finney, Paul Polakis and Al Smith in this work.

REFERENCES
Berzofsky, J.A. (1985) *Science* **229** 932–940.
Conti-Tronconi, B., Tzartos, S. & Lindstrom, J. (1981) *Biochemistry* **20** 2181–2191.
Crawford, G.D., Correa, L. & Salvaterra, P.M. (1982) *Proc. Natl. Acad. Sci. USA* **79** 7031–7035.
Dixit, V.M., Galvin, N.J., O'Rourke, K.M. & Frazier, W.A. (1986) *J. Biol. Chem.* **261** 1962–1968.
Djavadi-Ohaniance, L., Friguet, B. & Goldberg, M.E. (1984) *Biochemistry* **23** 97–104.

Dunn, S.D., Tozer, R.G., Antczak, D.F. & Heppel L.A. (1985) *J. Biol. Chem.* **260** 10418–10425.

Finney, K.G., Messer, J.L., DeWitt, D.L. & Wilson, J.E. (1984) *J. Biol. Chem.* **259** 8232–8237.

Hessova, Z., Thieleczek, R., Varsanyi, M., Falkenberg, F.W. & Heilmeyer, L.M.G., Jr. (1985) *J. Biol. Chem.* **260** 10111–10117.

Holowka, D., Conrad, D.H. & Baird, B. (1985) *Biochemistry* **24** 6260–6267.

Hopp, T.P. & Woods, K.R. (1981) *Proc. Natl. Acad. Sci. USA* **78** 3824–3828.

Jemmerson, R. & Paterson, Y. (1986) *Bio Techniques* **4** 18–31.

Kohler, G. & Milstein, C. (1975) *Nature (London)* **256** 495–497.

Lad, P.J., Schenk, D.B. & Leffert, H.L. (1984) *Arch. Biochem. Biophys.* **235** 589–595.

Lambin, P. & Fine, J.M. (1979) *Anal. Biochem.* **98** 160–168.

Lamy, J., Lamy. J., Billiald, P., Sizaret, P.-Y, Cave, G., Frank, J. & Motta, G. (1985) *Biochemistry* **24** 5532–5542.

Lillenhoj, H.-S., Choe, B.-K. & Rose, N.R. (1982) *Proc. Natl. Acad. Sci. USA* **79** 5061–5065.

Lo, M.M.S., Tsong, T.Y., Conrad, M.K., Strittmatter, S.M., Hester, L.D. & Snyder, S.H. (1984), *Nature (London)* **310** 792–794.

Makino, O., Shibata, Y., Maeda, H., Shibata, T. & Ando, T. (1985) *J. Biol. Chem.* **260** 15402–15405.

Mierendorf, R.C., Jr. & Dimond, R.L. (1983) *Anal. Biochem.* **135** 221–229.

Minitz, K.P. & Brimijoin, S. (1985) *J. Neurochem.* **45** 284–292.

Moore, G.R. & Williams, R.J.P. (1980) *Eur. J. Biochem.* **103** 543–550.

Polakis, P.G. & Wilson, J.E. (1984) *Arch. Biochem. Biophys.* **234** 341–352.

Polakis, P.G. & Wilson, J. E. (1985) *Arch. Biochem. Biophys.* **236** 328–337.

Rakonczay, Z. & Brimijoin, S. (1986) *J. Neurochem.* **46** 280–287.

Reznick, A.Z., Rosenfelder, L., Shpund, S. & Gershon, D. (1985) *Proc. Natl. Acad. Sci. USA* **82** 6114–6118.

Rockwell, P., Beasley, E. & Krakow, J.S. (1985) *Biochemistry* **24** 3240–3245.

Sams, C.F., Hemelt, V.B., Pinkerton, F.D., Schroepfer, G.J, Jr. & Matthews, K.S. (1985) *J. Biol. Chem.* **260** 1185–1190.

Santos, E., Tahara, S.M. & Kaback, H.R. (1985) *Biochemistry* **24** 3006–3011.

Schirch, D., Nemat-Gorgani, M. & Wilson, J.E. (1986) *Arch. Biochem. Biophys.*, in press.

Tainer, J.A., Getzoff, E.D., Alexander, H., Houghten, R.A., Olson, A.J., Lerner, R.A. & Hendrickson, W.A. (1984) *Nature (London)* **312** 127–134.

Tzartos, S.J., Rand, D.E., Einarson, B.L. & Lindstrom, J.M. (1981) *J. Biol. Chem.* **256** 8635–8645.

Ureta, T., Smith, A.D. & Wilson, J.E. (1986) *Arch. Biochem. Biophys.* **246** 419–427.

Vaidya, H.C., Dietzler, D.N. & Ladenson, J.H. (1985) *Hybridoma* **4** 271–276.

Vartio, T,. Zardi, L., Balza, E., Towbin, H. & Vaheri, A. (1982) *J. Immunol. Meth.* **55** 309–318.

Weldon, S.L., Mumby, M.C., Beavo, J.A. & Taylor, S.S. (1983) *J. Biol. Chem.* **258** 1129–1135.

Weldon, S.L. & Taylor, S.S. (1985) *J. Biol. Chem.* **260** 4203–4209.

Westhof, E., Altschuh, D., Moras, D., Bloomer, A.C., Mondragon, A., Klug, A. & Van Regenmortel, M.H.V. (1984) *Nature (London)* **311** 123–126.

Wilson, J.E. & Smith, A.D. (1984) *Anal. Biochem.* **143** 179–187.

Wilson, J.E. & Smith, A.D. (1985) *J. Biol. Chem.* **260** 12838–12843.

Wolf, C.R., Oesch, F., Timms, C., Guenther, T., Hartmann, R., Maruhn, M. & Burger, R. (1983) *FEBS Letters* **157** 271–276.

Index